David Levy's Guide to Observing Meteor Showers

Meteors occur when a meteoroid, a speck of dust in space, enters the Earth's atmosphere. The heat generated when this happens causes the surrounding air to glow, resulting in "shooting stars". During the most spectacular meteor storms, larger particles give rise to fireballs and firework-like displays!

Meteors are a delightful observing field – they do not require a telescope, and they can be seen on any clear night of the year, even in bright twilight. It was the sight of a single meteor that inspired David Levy to go into astronomy, and in this book he encourages readers to go outside and witness these wonderful events for themselves. This book is a step-by-step guide to observing meteors and meteor showers. Any necessary science is explained simply and in clearly understandable terms. This is a perfect introduction to observing meteors, and is ideal for both seasoned and budding astronomers.

DAVID H. LEVY is one of the most successful comet discoverers in history. He has discovered 22 comets, nine of them using his own backyard telescopes. Together with Eugene and Carolyn Shoemaker at the Palomar Observatory in California he discovered Shoemaker–Levy 9, the comet that collided with Jupiter in 1994 producing the most spectacular explosions ever witnessed in the Solar System. He is involved with the Jarnac Comet Survey, and is Science Editor for *Parade* magazine, and contributing editor for *Sky and Telescope*. He has been awarded five honorary doctorates, and asteroid 3673 (Levy) was named in his honor. His other recent books include *David Levy's Guide to Observing and Discovering Comets* and *David Levy's Guide to Variable Stars* (both Cambridge University Press).

Comet McNaught sets over a mountain range over Tucson, Arizona, a few hours before it rounded the Sun in January 2007. Photo by David Levy.

David Levy's Guide to Observing Meteor Showers

DAVID H. LEVY

CAMBRIDGE
UNIVERSITY PRESS

CAMBRIDGE UNIVERSITY PRESS
Cambridge, New York, Melbourne, Madrid, Cape Town, Singapore, São Paulo

Cambridge University Press
The Edinburgh Building, Cambridge CB2 8RU, UK

Published in the United States of America by Cambridge University Press, New York

www.cambridge.org
Information on this title: www.cambridge.org/9780521696913

First published 2008

Printed in the United Kingdom at the University Press, Cambridge

A catalogue record for this publication is available from the British Library

ISBN 978-0-521-69691-3 hardback

For Wendee, with loving and happy memories of our many nights under the stars. From 2164 Leonids during the storm of November 2001 to the occasional bright meteor soaring across the sky, we will never forget the many times we have observed meteors together.

Contents

Preface

Who has not, at some time or another, had a fresh new idea suddenly strike him and then wondered in amazement, "Why did I never think of that before?" One May evening, when I was fifteen years old, I was standing out in the front yard when just such a new thought came to me. The night air was soft and warm, there was no moon, and all the brighter stars were shining. Something – perhaps it was a meteor – caused me to look up for a moment. Then, literally out of that clear sky, I suddenly asked myself, "Why do I not know a single one of those stars?"

– Leslie C. Peltier, *Starlight Nights*, 1965.

Just after the end of World War 1, my mother and I were observing Perseid meteors when we saw a Perseid as bright as Venus. At the same instant, a brilliant sporadic meteor hurtled down at an angle 120 degrees from the Perseid's radiant. Both left sparks as they flew, and they appeared to collide. I never forgot that night.

– Dorrit Hoffleit, to David H. Levy, 1998

How does one develop a passion for the sky? Is it through careful mathematical study of the physical properties of the planets or cosmic distances throughout space? Or can it be through a simple, unintended look at the night sky thanks to a meteor? From the quotes above, it seems likely that for at least two of history's greatest astronomers, Dorrit Hoffleit of Yale University and comet discoverer and variable star observer Leslie Peltier, meteors were the trigger. As we shall see in Chapter 1, a single meteor that I witnessed fifty years ago inspired me to the lifelong passion I have for the sky. These nights I spend quite a bit of time each evening setting up several telescopes that search the sky for comets, but often my favorite sessions are the simplest – a hammock or lawn chair, a clear sky, and a meteor shower.

Meteor observing is a delightful observing field. It does not require a telescope, and it offers the unexpected chance of a surprise encounter with something from space. We enjoy the spectacle of a sky unconfined by eyepiece or dome slit. Because meteor observing requires almost no equipment, and only a basic knowledge of the constellations and magnitudes, meteors should be a most popular interest area for beginners. They are especially fun to watch as a team sport, conducted as groups assemble for shower parties for the most important meteor streams.

Meteors are teachers of the sky, for observing meteors involves learning some basic facts about where the constellations are, the magnitude differences between stars, and the relationship of solar system objects to the Earth.

More important, meteor showers give us a clear impression of the Earth pushing its way through space. As meteors stream from an apparent point in space, we get an illustration of the Earth moving along a railroad track through the Solar System. The meteors aren't really coming out from a single point; they are coming towards us in parallel paths; the appearance of a point is merely an effect of perspective that moves farther ahead as Earth proceeds along its annual trek around the Sun.

Through this book, I hope to inspire you to see the magic of meteor observing, whether you schedule an all night observing session, or are driving down a street and suddenly spot an exploding fireball lighting up the night. If your observing lasts eight hours or eight seconds, it should be fun, memorable, and educational. That's a big return from looking towards the heavens to spot a speck of dust.

Acknowledgements

In preparing this book, I am indebted to my wife Wendee, who prepared the index and whose enthusiastic encouragement led to the success of this entire project. I thank Tim Hunter and Brian Marsden for their many suggestions; Bobbi and Larry Gershon, Stephen James O'Meara, and Al Stern, who assisted with the chapter on the Leonids. My old friends at Twin Lake Camp might have noticed the same meteor that started my plunge into astronomy in 1956. My colleagues at Cambridge University Press did a fabulous job in producing this book. Finally, I thank my readers in the hope that this dip into astronomy will be a lasting and joyous one.

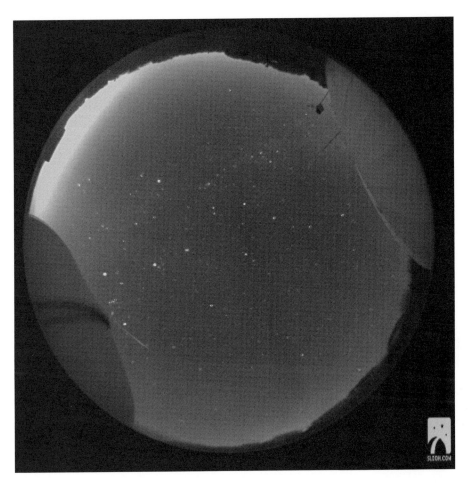

A meteor crosses through the field of the automated SLOOH telescope. Courtesy of and thanks to Tierney O'Dea, SLOOH.com.

1

July 4, 1956

Twin Lake Camp

Catch a falling star and put it in your pocket
Never let it fade away.
Catch a falling star and put it in your pocket
Save it for a rainy day.[1]

As a spindly asthmatic eight-year-old Canadian boy, I was not familiar with the Fourth of July, 1956, as a national holiday. Four days earlier, I had boarded a train in Montreal with my two brothers Richard and Gerry and my sister Joyce. It was the first time in my life I would be away from Mom and Dad, and I was not happy about it. I was homesick before the train pulled to a stop by the small station at Whitehall in New York state. We boarded buses – one for Joyce who would attend nearby Camp Awanee, and another for the rest of us. Richard was looking forward to his elite experience on Senior Hill; Gerry was wondering about life in Freshman House, and I was headed for Bunk B on Junior Row. With these thoughts in our minds, our buses crossed the border into Vermont, near the historic Fort Ticonderoga where American soldiers fought the British during their war of independence. We reached Vermont Route 30 and headed south through beautiful undulating mountain scenery. Finally we passed a restaurant on the right called The White House, descended a hill, slowed down, then turned left. We had reached Twin Lake. (See Figure 1.1.)

I didn't enjoy that first night. My first request of the Camp Director, my one lifeline to big city life, was to be allowed to watch the Perry Como show each

---

[1] Perry Como, 1957.

Figure 1.1 A group of children at Twin Lake Camp in the Summer of 1956. The thin boy second from right is the author, then eight years old. He saw his first meteor that summer.

week. "If I let you do that, David, then I'd have to let 160 other boys watch their favorite shows from my living room!" he explained. Yielding to his logic, I returned to Bunk B where I tossed and turned. I walked onto the porch and joined the counselors, who pointed out the mountain peaks and how they appeared to interact with the cloudy sky. I was surprised by how dark it was. It was my first night away from the bright lights of Montreal, and my first night away from television shows like Perry Como. But they showed me how the country had a new kind of magic.

Bunk B had five children, and we all quickly became good friends. Our first major celebration at camp took place four days into the season, on July 4. The evening began with an event in the social hall. Then we headed down to an open field to enjoy the camp's fireworks display. When it ended, we little guys were sent back to our bunks. The way back to Bunk B was a slow uphill climb toward the north. As we walked casually up the gentle slope, my gaze turned skyward. Toward the right, a bright star highlighted the east. I was amazed at how many stars there were. I stared for a few seconds. Then a "falling star" – a meteor – streaked across a small portion of sky. It wasn't particularly bright – perhaps as bright as Polaris, the north star, and it did not leave a long trail. It moved fairly slowly and headed toward that bright star that was almost certainly Vega. It lasted only about a second.

To germinate and sprout

That's the story – except that that one-second event planted a seed in my mind. It didn't germinate for about 15 months. I do remember walking home from school on an early October afternoon in 1957, and one of my classmates told me that there was a new moon in the sky called *Sputnik*. I looked up into the cirrusy afternoon sky and wondered if I could see this remarkable thing there and then. I was grabbed by the wonder of the new moon, but was too young to share the concern my parents felt over what it represented that evening at dinner.

In the spring of 1960, the seed finally sprouted. On June 11, I started writing a book called *A look at the stars*. (I doubt I finished it.) On June 21, while cycling to graduation and a picnic at Roslyn School, I fell off my bicycle and broke my arm. My cousin, Roy Kaufman, gave me a book called *Our Sun and the Worlds Around It*. It was filled with pictures and good information; I used that book as a basis for writing *The Wonders of the Earth*, which was about our Solar System. By the end of the summer of 1960, the night sky had become an all-consuming passion which culminated on the evening of September 1, 1960, when my parents and I set up my new 3.5-inch reflector telescope and turned it toward the brightest thing in the sky. As I focused the telescope, the image sharpened to a bright disk with two dark bands crossing it, and four little stars nearby. From a meteor to Jupiter, I was now a lover of the night sky.

Dating the meteor

So many memories of youth get lost over the years, but not the meteor incident. I recall precisely where at the camp I was when I saw it. I also remember that it followed a Fourth of July celebration. But I spent three summers at Twin Lake; which year was it? I assumed it was my final year there, 1958, when at age 10 the incident would have meant the most to me. I did remember a very dark night, something that would be easy to confirm using a planetarium program that displays the sky, with proper position and phase of the Moon, on any night. When it showed the sky from central Vermont on July 4, 1958, a bright waning gibbous Moon rose less than an hour after dark. By the time the social evening and fireworks were over, I would have been observing a sky not much darker than a Montreal sky. The previous summer was not much better: On July 4, 1957, an 8-day old Moon next to Spica and slightly-past first quarter brightened the southwestern sky.

On the evening of July 4, 1956, the waning crescent Moon was within three days of new, and not in the sky.

But was it clear? What about what the weather was gonna do, would the sky be clear? What was the weather on July 4 of each of the three summers I was at camp? For answers I turned to two sources: the National Weather Service, and John Martin, senior systems analyst at Newsbank, a company headquartered in Chester, Vermont. Both the Weather Service and John, through his access to newspaper archives, confirmed that Independence Day 1958 was cloudy with rain, as part of a long-lasting cold system that, I recall, seriously aggravated my asthma that summer. The 1957 evening had variable clouds with thunder-showers scattered throughout the region. The National Weather Service, who provided me with weather data for those three nights, reported that on that evening the sky was clear and dry throughout Vermont. July 4, 1956, was the night I saw my first meteor.

A seed was planted that night, but it took some time to germinate and then to sprout. I had no idea at the time that what I had witnessed was the end of a 4.5-billion year life story. I'd later learn that my falling star was actually a speck of dust from the Omicron Draconid stream, which might have begun its life as part of Comet Metcalf, but those details will come later in Chapter 10, after we have some more background about meteors. For now, what counts is that after wandering across the Solar System, past the Sun, planets, moons, comets, and asteroids, this tiny particle separated from the comet and began a life of its own. Exploring space by itself now, the particle moved through the Solar System for hundreds of thousands of years. In the summer of 1955 the Earth was a bright star in the particle's sky, by the spring of 1956 the Earth was getting bigger by the day. Throughout the day on July 4, the Earth loomed larger as nightfall approached the eastern United States. At last the particle slammed into Earth's atmosphere at a speed of 14.7 miles/s (23.7 km/s).[2] It heated the surrounding air until it was incandescent. Then it was gone.

I am writing this chapter on July 4, 2006, precisely 50 years after that meteor. A lot has happened in those 50 years. By 1960 I was completely hooked on astronomy. All the comets I have found, all the eclipses I have seen, all the young people who have looked through my telescopes – all those things can be traced back to that one simple event, a meteor falling from a clear, dark sky while I stood in a field with some friends at camp.

[2] Coincidentally the precise velocity required to escape from Neptune!

2

What is a meteor?

Never tell me that not one star of all
That slip from heaven at night and softly fall
Has been picked up with stones to build a wall.
Some laborer found one faded and stone cold ...[1]

A meteor is actually not an object but an event that occurs when a meteoroid, or a speck of dust, makes contact with Earth's atmosphere and heats the surrounding air to incandescence. But if the physics behind this sand-grain-sized speck of dust is complicated – its orbit around the Sun, and how it can disappear in an instant after wandering through the Solar System for billions of years – then the idea of a meteor brightening the sky for a second or two is something an inquiring child can grasp. And for that child, the first step towards understanding is to appreciate that after each meteor falls there are still just as many visible stars as there were before. Every star is a sun, in some ways like our own. These distant suns will not fall to Earth. A meteor takes place close to us, perhaps 40 miles above us in the Earth's own upper atmosphere.

So, once again: a *meteor* is an *event* that takes place when a tiny particle, called a meteoroid, enters the Earth's upper atmosphere, heating the air around it to incandescence. It is that incandescent glow that we see as a meteor. A *meteoroid* is the particle that produces a meteor. It starts out as a particle of dust in a comet or, less often, in an asteroid. The particle continues circling the Sun in roughly the same orbit as the comet from which it came, but over time it gets farther from that comet, whose orbit becomes littered with these particles. See Figure 2.1.

---

[1] Robert Frost, A Star in a Stone-Boat, *Robert Frost: Collected Poems, Prose, and Plays*, eds. Richard Poirier and Mark Richardson (New York: Library of America, 1995), p. 162.

Figure 2.1 Comet McNaught at the moment of perihelion, its closest approach to the Sun. This photograph was taken at noon, with the comet visible in broad daylight near the Sun.

The meteor event

A meteoroid's final plunge begins somewhere from a height of 75 to 50 miles, and can vanish at altitudes between 50 and 40 miles. Its velocity depends on whether it is catching up from behind or hitting the planet head on. If the meteoroid is directly behind Earth, it will strike at about 7.5 miles per second; if the collision is directly head-on, its velocity can reach 45 miles per second. Usually the velocity will be somewhere in between those extremes. Some meteors show a trail, a streak lasting a second or two. Rarely a bright meteor will leave a train of faint light that can last for several minutes; I once saw a Taurid so bright that dogs were wakened and started barking, and its train lasted for 20 minutes.

Micrometeoroids and Brownlee particles

There are particles even smaller than meteoroids. These particles, called micrometeoroids or *Brownlee particles* (after Donald Brownlee who first described them) are tiny cometary particles from the zodiacal light cloud, some of which are from ancient comets, and actually arrive on Earth. As small as these particles are, micrometeorites are actually easy to spot on a dark night. In northern hemisphere winter, they appear best as a tepee-shaped cone of light, called the *zodiacal light*, which climbs sharply up from the western horizon. In the southern hemisphere the zodiacal light climbs high in the sky on June, July, and August mornings. Experienced observers under very dark conditions can follow the zodiacal light all around the sky. As the tepee shaped zodiacal light fades, it continues as a thin, faint *zodiacal band* that follows the ecliptic across the sky. At the point in the sky opposite the Sun, it widens out into the *Gegenschein*, or counterglow.

In 1970 the first balloon experiment to collect dust particles in the stratosphere retrieved artificially produced and volcanic dust as well as some microscopic samples of feathery cometary dust. These Brownlee particles are extraterrestrial in origin. Specially equipped planes flying some 20 kilometers into the stratosphere routinely collect such particles today. The Brownlee particles eventually drop to the Earth's surface at the approximate rate of a single particle on each square meter of the Earth's surface every day.[2]

In the early years of the Solar System, the zodiacal dust was far more extensive than it is now. Over a long period of time, uncountable billions of such grains entered the atmospheres of Earth, Mercury, Venus, Mars, and struck the airless Moon. They were not large enough to be consumed as the meteors we see in the sky at night; instead they were so tiny that they came to a stop in the upper atmosphere and then wafted gently down to the planetary surfaces. Today, lesser numbers of these primordial grains reach the surface. Occasionally a micrometeorite may form the nucleus of a raindrop.

Fireballs and bolides

If a meteor reaches or exceeds the brightness of Venus, it is called a *fireball*. Sometimes a fireball explodes as it falls; if you see a bright meteor break up in a shower of sparks, then you have seen a *bolide*. A fireball brighter than the quarter Moon might actually survive its plunge through the atmosphere to land on Earth's surface. Listen for rumbling or detonation sounds after you witness such a thrilling event. In September 1968 I saw such a meteor fall towards the northeast. It lasted about half a minute, suffered several explosions, and finally disappeared over the northeastern horizon. Several minutes later I

---

[2] M. Bailey, S. Clube, and W. Napier, *The Origin of Comets* (Oxford: Pergamon Press, 1990), pp. 454–455.

heard a low rumbling noise that lasted almost half a minute. It is possible that this object survived its fall to Earth, and that the meteorite still awaits discovery in the forests of Prince Edward Island. But to have any chance of recovering a meteorite, accurate observations of the altitude and azimuth of the beginning and end points of the fireball are necessary from several sites. From these measurements, one can try to calculate where the meteorite landed.

Meteors are more numerous on certain nights

Although meteors can be seen on any night of the year, at certain times of the year the Earth rushes through a stream of particles that are strung along the orbit of a comet. Kicking off the new year are the Quadrantids, which are active for only a few hours on the night of January 2/3. The Lyrids peak at the end of April. We call this shower the Lyrids because its meteors appear to radiate from the constellation of Lyra. The Eta Aquarids, from Halley's comet, are active around the fourth of May.

The next good shower is the Delta Aquarids, which reaches its maximum on July 29 with meteors one could trace back to a point in the sky near the star Delta in the constellation of Aquarius. By itself, the Delta Aquarids is only a moderate shower, but its maximum marks the onset of one of the northern hemisphere's great showers, the Perseids. They peak two weeks later on August 12. The next major shower, the Orionids, follows on October 20, and, like the Eta Aquarids of May, the Orionids derive from Halley's comet. The Taurids, which can display some dramatic bright meteors, are visible during the first part of November. The Leonids occur on November 17, and the Geminids reach their strong maximum on December 13.

Why do meteors of a particular shower appear to radiate from a single point in the sky? The effect is due to perspective, and is similar to looking down a railroad track. The tracks appear to converge at a specific point, but they obviously can't, since the tracks must be parallel or the trains will all fall off. Similarly, the meteors from a particular shower appear to be coming from a specific place in the sky. But they actually hit Earth from the *same* direction – they are on parallel paths.

A meteor shower radiant is actually a small area, the size of which indicates the age of the stream. A small radiant size indicates a youthful, hardly dispersed stream. A few meteor showers are complexes that show several radiants a few degrees apart in the sky. Because the Earth moves relative to the meteoroid orbits, the apparent radiant shifts from night to night. Not all meteors belong to showers; *sporadic* meteors are thought to be the remains of ancient streams that have had their orbits dispersed beyond recognition.

More meteors are seen in the hours before dawn than after sunset. This effect occurs because in the morning hours meteoroids are colliding more nearly head-on, rather than catching up with Earth as they do in the evening hours. The closer to the zenith the radiant is, the higher the meteor rate usually will be.

Amateur meteor observing

In the early 1920s Charles P. Olivier founded the American Meteor Society (AMS), a group of dedicated meteor observers whose observations would build up a central archival file of meteor work done over many years. Dr. Olivier endorsed a mode of operation that concentrated on the efforts of single observers and not with group reports. In the 1950s, when planning for the International Geophysical Year (IGY), large numbers of meteor observations were accumulated. This international program of cooperation between countries included professional and amateur astronomers alike. In no country was this program taken more seriously than Canada, where Dr. Peter Millman of the National Research Council carefully prepared an observing form that would be a model of simplicity and could be understood by almost anyone. For each meteor, only two pieces of data were needed: magnitude and shower membership. The time column was left unlined, so that observers could record time to the hour, minute, or second. Here was an "observer-friendly" program that, if successful, could attract hundreds of Canadian amateur astronomers. These forms are used in this book.

From the time the program began at the dawn of the IGY in 1957 to its official close almost fifteen years later, groups of observers across the country organized major observing projects for most of the major annual showers. The observing became so intense that for some astronomy societies in Canada, meteor observing was the major observational activity. From years of observing by the AMS, IGY, and other groups, amateurs have achieved a monumental file of archival data that shows the changing strength of meteor showers over three-fourths of a century. The data required are easily obtained and provide both the novice and the experienced observer with an opportunity to contribute to the exciting field of meteor research.

3

Some historical notes

Goe, and catche a falling starre,
Get with child a mandrake roote,
Tell me, where all past yeares are,
Or who cleft the Devil's foot.
Teach me to heare Mermaides singing,
Or to keep off envies stinging,
          And find
          What winde
Serves to advance an honest minde.[1]

I would more easily believe that two Yankee professors would lie, than that stones
would fall from heaven.[2]

Catch a falling star, and put it in your pocket,
Save it for a rainy day.[3]

Go and catch a falling star. This four-century-old piece of advice was
modernized in the late 1950s: go outdoors at night, look at the stars, and catch
a meteor. As unbelievable as Thomas Jefferson claimed it was, stones do fall
from heaven. What you catch are not real stones but the excitement of that
flash of light that brightens the sky and makes you wonder. When you see a
meteor, you share an experience with many, many people throughout history.
Meteors are often mentioned in the works of Shakespeare, John Donne, and
other writers from many times.

---

[1] John Donne, Go and catch a falling star, in *Poems and Prose* (New York: Everyman's Library of
   Alfred Knopf, 1995), p. 14.
[2] Thomas Jefferson, 1807.   [3] Perry Como, 1957.

Ancient observations

The history of meteors and meteorites goes back thousands of years. Iron meteorites played a crucial role in the dawn of the Iron Age more than 3000 years ago. At different times in different geographical areas, cultures used iron from meteorites as part of their supply to make tools or weapons.

There is a story about Constantine crossing the Alps in preparation for the battle against the Emperor Maxentius in October 312. The battle took place at the Milvian Bridge, a stone structure crossing the River Tiber. If the account of battle told decades later by Eusebius is correct, Constantine saw, or was told of, a bright light in the daytime sky that he described as a blazing cross. Could this event have been a magnificent bolide exploding in the sky, its fragments spreading out in the form of a cross? The event inspired Constantine to go on and win the battle as a Christian. If this interpretation is correct, then a bright meteor profoundly changed the history of Rome and the history of Christianity.

In some portions of history, trees talk better than humans. In 536 CE, writers complained about the Sun's light being dimmed, as if in eclipse, for an entire year. Fruits went unripened, and, perhaps as a result of these events, the plague spread through many lands. Dr. Michael Baillie of Queen's University, Belfast, noted that these historical events could be corroborated by thinner tree rings. With all this evidence pointing to diminished sunlight for an extended period, is it possible that the impact of a large meteorite triggered these events? We do know, from theoretical modeling, from studies of rocks laid down in the wake of the impact that destroyed most life on Earth 65 million years ago, and from observations of the impacts of Comet Shoemaker–Levy 9 on Jupiter in 1994, that effects like these follow impacts of objects from space onto the surface of the Earth.

On November 7, 1492, thousands of people witnessed a bright fireball fly over the Rhine valley. A few minutes later a 12-year-old boy saw a rock drop into a field. He found a rock that had dug itself into a small crater. That 12-year-old made the first historical recovery of a meteorite after a fall. It was a long-awaited idea that stones could fall from the sky that dates back to Diogenes, the philosopher of Apollonia in the fifth century BCE. The idea was not taken seriously for more than 2000 years after that. In 1772, Peter Pallas, a German scientist, studied a very large rock near Krasnojarsk, Siberia. Pallas reported on this large and rare sample of rock, which prompted another German scientist, Ernst Florenz Chladni, to revive Diogenes' theory. In 1794 he wrote that meteorites come from space, and could be related to bright fireballs that appear from time to time in the sky. For the next decade most scientists ridiculed this theory, but on April 26, 1803, thousands of small stones fell on the French village of L'Aigue. The fall was witnessed by most of the population of the town that day, and definitely turned theory into reality.

The Leonids of 1833

If the nineteenth century opened with an improvement in theory, it continued with a vast enhancement in observation. In November 1832, and again in November 1833, hundreds of thousands of meteors blanketed the skies over Europe and America. It was not the first time such showers were visible; during the time John Donne was writing the words that open this chapter, there were reports in the Chinese text *Thien-Wen-Chih* that on November 6, 1602, "Hundreds of large and small stars flew, crossing each other." The Korean text *Munhon-Piko* notes that on November 11, 1602, "Many stars flew in all directions." One of the texts probably had an error in date; it would seem that only on one of those dates would there have been a major meteor storm in the Far East. Could Donne have seen this storm? If the storm was visible in China it was probably not visible on the other side of the world in England, since meteor storms, as opposed to showers, do not last more than a few hours. In any event there do not appear to be any records that the storm was seen in Europe.

In 1866 another meteor storm occurred, and that time the meteors were connected to Comet Swift–Tuttle's 33-year orbit about the Sun. The fact that these storms appeared with such regularity encouraged the early meteor scientists to predict another storm in 1899.

Decade of the parents: meteor showers and comets

While bright meteors and greater meteor showers have been chronicled for millennia, the Leonids of 1833 is widely considered to be the spark that ignited serious study of meteors. Only 30 years later, the 1860s saw a fortuitous series of comets, discovered one after another, whose orbits closely matched those of some of the major showers. The first comet to reveal its dusty trail was Swift–Tuttle. Discovered in July 1862 by Lewis Swift from the sleepy town of Marathon in northwestern New York, and later at the Harvard College Observtory by Horace Tuttle, the comet became as bright as the North Star, easily visible to the unaided eye. "The last time I saw the comet," co-discoverer Tuttle wrote, "was on the night of the 17th of September, after I had entered the federal army and gone to camp. It was then about six degrees [about 12 diameters of the full Moon] from Antares and just visible to the naked eye."[4]

In 1866 Tuttle discovered another comet, sharing this find with Ernst Tempel. That same year Giovanni Schiaparelli, an astronomer later famous for

---

[4] H.P. Tuttle, Schreiben des Herrn Tuttle an den Herausgeber, *Astronomische Nachrichten* **59**, no. 1404 (1863), vols. 187–190.

Figure 3.1 Leonid fireball near Aldebaran. Photo by Tim Hunter and James McGaha.

his theory of canals on Mars, proposed a remarkable idea that meteor streams are the residue from comets. As examples he showed that the orbit of the famous Perseid meteor shower was analogous to that of Swift–Tuttle, and that newly discovered Tempel–Tuttle is the parent of the Leonid meteors. See Figure 3.1.

So not only did Swift, Tempel, and Tuttle have bright comets to their credit, but they also had the one whose debris would fall in a meteor shower every August.

What can meteors teach us about their parent comets?

As a graduate student at the University of California at Berkeley in 1930, Fred Lawrence Whipple was part of a team that calculated a rough orbit for the newly discovered planet Pluto. By the end of World War II, Whipple was at Harvard University. Still using his skill in computing orbits by hand, Whipple made an unexpected discovery. The orbit of each meteor should be in a virtually unchanging ellipse. But the orbits were changing with time; the meteors traveling in them were slowly approaching the Sun. If this process has been happening since the Solar System's birth, then there should no longer be any meteors in the sky at all.

Whipple struggled to understand this mystery and to solve it. Since the mid-nineteenth century scientists have understood that meteors traveled in the same orbit as their parent comets. Comets were then thought to be huge "flying sandbanks." Whipple decided that if comets were indeed sandy, they would not provide a present-day source for the meteors. To escape this dilemma, Whipple redefined our understanding of what makes up a comet. He built this model on the basis of Encke's comet, which returns to the Sun's vicinity very often – every three years. Comets, Whipple proposed, are large conglomerates of ices and meteoric particles, now popularly known as "dirty snowballs."[5] In September 1951 Fred Whipple published the results of his work using the orbit of Encke's Comet in two of the most important papers in the history of comet science.[6]

Back to meteor storms

Full meteor storms are extremely rare, but observers should always be ready. In 1846, Comet P/Biela split into two comets, and was observed as a pair of comets in 1852. It was poorly placed during its 1859 apparition, but the next one, in 1865–66, should have been favorable. However, no trace of it could be found. Then, at the time of its projected return in 1872, a storm of meteors appeared, described thus by Gerard Manley Hopkins, a famous English Victorian Poet and observer:

> Great fall of stars, identified with Biela's Comet. They radiated from Perseus or Andromeda and in falling, at least I noticed it of those falling at all southwards, took a pitch to the left half-way through their flight. The kitchen boys came running with a great todo to say something redhot had struck the meatsafe over the scullery door with a great noise and falling into the yard gone into several pieces. No authentic fragment was found but Br. Hostage saw marks of burning on the safe and the slightest of dints as if made by a soft body, so that if anything fell it was probably a body of gas, Fr. Perry thought. It did not appear easy to give any other explanation than a meteoric one.[7]

---

[5] F.L. Whipple, *The Mystery of Comets* (Washington: Smithsonian Institution Press, 1985), pp. 145–147.

[6] Whipple announced his theory in two papers. A Comet Model. I. The Acceleration of Comet Encke, *Astrophysical Journal* **111** (1950), 375–394, explains how the orbit of Periodic Comet Encke, which is shrinking with each return, is interpreted if the structure of its nucleus consists of meteoric material embedded in ices which sublimate to gases. The freed material rushing out of the comet with some force can accelerate the comet. The second paper, Physical Relations for Comets and Meteors, *Astrophysical Journal* **113** (1951), 464–474, expands on this model.

[7] G.M. Hopkins, *The Journals and Papers of Gerard Manley Hopkins*, eds. H. House and G. Storey (London: Oxford University Press, 1959), pp. 227–228.

As the years moved forward, plans continued to observe a much anticipated Leonid meteor storm in 1899. Two years before that, explorer Robert Peary learned of a great and unusual rock in Greenland that he assumed must be a big meteorite. In 1897 he used hydraulic jacks and rails to load the meteorite onto a ship, then sailed to New York. That October, it was moved to the museum by a 28-horse team coming along Broadway. The huge meteorite remains at the American Museum of Natural History, where my family and I saw it in 1959.

Two years after the meteorite arrived in New York, meteor watchers prepared for the meteor storm of 1899. It never came. The meteor counts were higher that year but nowhere near the levels of 1866, and in 1933, they were even lower. We continue the specific story of the Leonid meteor shower, and how its activity changed during the twentieth century, in Chapter 16. For now, we return to the turn of the twentieth century, and to Grove Carl Gilbert, one of the top geologists in the country at the time. He traveled to northern Arizona, where, near the tiny town of Winslow, a huge crater sat. Gilbert was the first to study the crater scientifically. He thought that the crater was the result of an iron-rich asteroid impact, but his tests didn't confirm that idea. He hoped that a huge mass of iron beneath the crater floor would cause an anomaly in his compass reading. It didn't, so he assumed that if the crater was an impact feature, then the object must be too far below the surface to cause a magnetic blip. He described the failure of his test in his retiring presidential address at the Geological Society of Washington, but from a philosophical point of view.[8] The paper discussed how a hypothesis is born, figured through, and then is either accepted or discarded. Meteor Crater (as the crater is now called) served the purpose of an example. "The mental process by which hypotheses are suggested," he wrote, "is obscure. Ordinarily they flash into consciousness without pre-monition." Gilbert proposed the idea that hypotheses come out of analogy. To explain an unusual event or a peculiar feature, the scientist looks to see what aspects of it might have been explained before, and builds the hypothesis on that.

Gilbert's paper addressed the two likeliest ways a crater is formed: by the explosive eruption of a volcano, or by the impact of an object from space. For craters here on Earth as well as on the Moon and other planets and moons, scientists are forever debating which giant pockmark was caused by one and which by the other; in fact Gilbert gave the first clear exposition that lunar craters could have been formed by impacts of other objects in space. "What would result," Gilbert now asked, "if another small star should now be added to

---

[8] G.K. Gilbert, *The Origin of Hypotheses: Illustrated by the Discussion of a Topographic Problem*, Presidential Address, The Geological Society of Washington, March 1896. See also *Science*, **3** (1896), 1.

Figure 3.2 Meteorite ALH84001, from Mars. Photo courtesy NASA's Johnson Space Center.

the Earth, and one of the consequences which had occurred to me was the formation of a crater?" Gilbert thought that the main body would be composed of iron, and that if it lay beneath the crater it should cause "a local deflection of the magnetic needle."[9] He noticed that the crater 45 miles east of Flagstaff was not circular but squarish. Perhaps the star "struck the earth and bounded off, finally coming to rest at some point further east?" But he found that the crater's shape was inconsistent with the asteroid hitting and then ricocheting (see Figure 3.2).

Gilbert concluded that the crater was formed by a steam explosion, like that at Mt. Vesuvius in AD 79 or to Krakatoa in 1883. After all, the crater is less than ten miles from a known volcanic crater. But – and Gilbert noticed this too – most volcanic craters are atop their mountains; this one bore only a casual relationship to the land around it. And what about the nearby meteorites? Could the meteorite fall have "touched the volcanic button," setting off a volcanic explosion?

In 1903 Daniel Moreau Barringer, a successful mining engineer and lawyer, heard about the crater and the platinum-laden meteorites surrounding it. He

[9] Gilbert: *The Origin of Hypotheses*, p. 11.

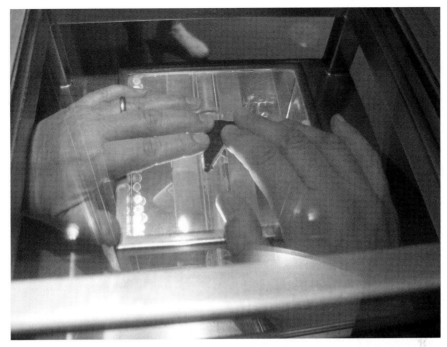

Figure 3.3 Wendee and David Levy touching a Moon rock at the Johnson Space Center, May 2006. Photo by David Levy.

believed that the main body of the meteorite was still hidden under the crater. Much of the rest of his life was spent in a search for this main mass of iron. The government later established a post office in the area and named it Meteor, named not for the crater but for the meteorites surrounding it.

Daniel Barringer's dream was never fulfilled during his lifetime. That would await Eugene Shoemaker, the man who would finally prove that Meteor Crater was the site of a big impact from space. He succeeded in discovering coesite, a form of quartz that is formed under temperatures so high not even a volcanic explosion would explain it. Only an asteroid striking the Earth at dozens of miles per second would accomplish that.

Over the last two centuries, scientists, astute amateur observers, and others have witnessed objects falling from the sky, from tiny specks of dust to the 50-yard-wide asteroid that formed Meteor Crater in a few seconds some 50 000 years ago. If we put all these together, we must conclude that we are a target of objects from space. Most of them are tiny, but some are large enough to be dangerous. See Figure 3.3.

4

Small rocks and dust in space

Captain: 'Tis thought the king is dead; we will not stay.
The bay-trees in our country are all wither'd
And meteors fright the fixed stars of heaven;
The pale-faced moon looks bloody on the earth
And lean-look'd prophets whisper fearful change... [1]

Isn't it strange that of all the many pieces of our Solar System, it is the smallest ones – comets, meteors, and an eclipse of the Moon, that attracted the attention of civilizations right up to Shakespeare's time 400 years ago? While the great worlds orbit the Sun, their motions are predictable and are not a cause for alarm. But a meteor shower did attract notice, along with an increasing interest in the workings of our Solar System. Our solar neighborhood is a big place, with several different types of object in its domain, ranging in size from one star, the Sun, to tiny micrometeoroids that are even smaller than the meteors this book is about. These objects include the Sun, planets, moons, comets, asteroids, meteoroids, and micrometeoroids. It seems a simple division, but recent debates tell us that the boundaries among these types of objects are fuzzy at best. See Figure 4.1.

Categorizing objects in our Solar System

The best known debate concerns the definition of a planet. In 2006 the International Astronomical Union defined a planet in a most controversial way. I would maintain that the Solar System, as we understand it in 2007, consists of 11 planets – Mercury, Venus, Earth, Mars, Ceres, Jupiter, Saturn, Uranus,

[1] William Shakespeare, *Richard II* 2.4.7–11. All Shakespeare references in this book come from the Tudor Edition, ed. Peter Alexander (London and Glasgow: Collins, 1964).

Figure 4.1 Tim Hunter took this photograph of the satellite Echo 1, in 1960. Vega and Lyra are at right.

Neptune, Pluto, and Eris, with perhaps several dozen more bodies on the way. These worlds are large enough to be spherical in shape – essentially bodies larger than about 800 kilometers in diameter. This definition was, in fact, what the IAU's own committee passed just before its General Assembly (GA) in 2006. At the GA that summer, however, the members added that a planet also has to "clear the neighborhood" around its orbit. This part of the definition, which rules out Pluto, Ceres, and Eris, is seriously flawed. Objects cross the orbits of all the planets except possibly Mercury and Venus. So if this definition is taken literally, even the Earth is no longer a planet!

If astronomers cannot agree on how the largest objects in the Solar System are categorized, how about the smaller objects? At some level, it is hard even to differentiate between a comet and an asteroid. In 1990 Carolyn and Gene Shoemaker and I discovered an asteroid, 1990 UL3, through the 18-inch Schmidt Camera at Palomar Observatory. Through that telescope the object looked like a star, hence the categorization of asteroid. But a few weeks later, astronomer Steve Larson and I, using a much larger 61-inch reflector and CCD, detected a coma surrounding this "asteroid" and a long, faint tail stretching from it. Hence the object became known as Comet Shoemaker–Levy 2, or P/1990 UL3.

As comets like Shoemaker–Levy 2 orbit the Sun, they shed dusty material that continues to circle the Sun as meteoroids or micrometeroids. A meteoroid can be as small as a speck of dust, or larger than a basketball. Objects smaller than a tenth of a millimeter in diameter are called micrometeoroids. Just like the hotly debated boundary between what is a planet and what isn't, the boundary between meteoroids and micrometeoroids is a fuzzy one.

Micrometeoroids and Brownlee particles

As small as they are, micrometeoroids are critically important to the evolution of our Solar System. The Apollo missions explored the Moon and its mountains, valleys, and craters. The astronauts discovered that while weathering and erosion are the dominant processes that sculpt the mountains on Earth, on the Moon, which has no weather, frequent strikes by micrometeorites are responsible for reshaping the Moon's surface appearance.

Particles this small are also called interplanetary dust particles or IDPs, and Brownlee particles, since it was Don Brownlee who demonstrated their extraterrestrial, cometary nature. They range in size from 5 to 50 microns (a micron is a millionth of a meter). Put together, these particles are easily seen on clear dark nights. Especially in February and March, they appear collectively in the northern hemisphere evening sky as a tall pyramid-shaped structure called the zodiacal light. It is a soft, white glow that is widest at the horizon, and peaks high in the sky along the ecliptic. If the sky is very dark and your vision really good, the glow continues as a band all along the ecliptic; it is called the zodiacal band. Around the point opposite the Sun, the band widens into a large, faint glow called the *Gegenschein*. It is best seen when it is not near the Milky Way, as in the months of September to November, and in February. It consists of dust grains in the plane of the Solar System, seen by reflected sunlight. These grains of dust come from comets as they cruise by the Sun on their way through the Solar System. The particles are small, perhaps 500 microns in diameter, and are separated from each other by about five miles.

I saw the Gegenschein for the first time on August 20, 1966, from a dark sky site in Vermont. The eerie oval-shaped glow, covering some 10–15 degrees (20–30 moons) of sky, was surreal; just a patch of faint haze hanging in the sky.

Micrometeoroids are so small that when they enter the Earth's atmosphere they do not heat to incandescence. Instead they slow down gradually and waft downwards through the atmosphere. A precious few found their way to some of the Gemini missions during the mid 1960s, and to high-flying planes like U2s, and are still being collected by NASA jets. These particles find their way to the surface of the Earth, at the approximate rate of one particle per square meter per day.

Fireballs and meteorites

At the opposite end of what falls from the sky are objects so big that they actually survive their fall through the atmosphere. If you see a meteor at magnitude –8 or brighter (about the brightness of the quarter Moon), there is a chance that it will hit the Earth; objects that survive and hit the Earth are called meteorites. They can land anywhere on Earth, including, as happened in 1992, on a car.

The Peekskill meteorite

On a clear Friday evening that October, thousands of people were outside watching high school football games all over the United States. In the town of Ashland, Ohio, David Hartsel, a well-known amateur astronomer, was enjoying the Ashland High football game with his children, Heather and Sean. Suddenly a yellow–green fireball emerged out of the twilight high in the southwest. Bright as a first quarter Moon, it raced across more than a sixth of the sky within the next ten seconds, leaving a bright trail and breaking up into small pieces. On the field, players and referees on both teams pointed to it, and everyone in the west stands saw the meteor fade as it approached the northeast horizon. It was the brightest meteor Hartsel had ever seen.

Hundreds of miles to the east, Michelle Knapp, a Peekskill, New York, high school senior heard a crashing sound outside her window. She found her car's trunk crushed and a warm rock the size of a football next to the gas tank. She reported what she thought was vandalism to the police. But the rock's warmth and heavy weight made the police wonder how anyone could have tossed it. They contacted a geologist, William Menke, who identified the rock as a stony chondrite meteorite with a high iron content. I suspect that other pieces of the 1992 meteorite still lie in the forest around Peekskill.

The meaning of the meteorite from Mars

Meteorites fall every day, but actually finding one is rare. The easiest place to find them is on a field of ice, for example in Antarctica. In 1984, a team of researchers found a large meteorite on the ice at the foothills of the Alan Hills, thus its label ALH84001. Studies of air pockets inside the meteorite revealed that the "air" inside matched what the *Viking* spacecrafts had detected on Mars in 1976. This was a meteorite that was shot out of Mars by a comet or asteroid impact, then orbited the Sun until it reached Earth, where it fell in the

Antarctic ice. Further research on this meteorite indicated that it could hold evidence of ancient life on Mars.

This rock probably hardened from magma as part of the original crust that formed when Mars cooled. The rock is so old that it can tell us directly about the period of late heavy bombardment that sent objects slamming into all the inner planets of the Solar System. During its time on Mars, this rock led an eventful life. One impact fortuitously happened close enough to the rock to crack it. The crash took place between 4.0 and 3.8 billion years ago, during the period of "late heavy bombardment" when the inner planets were subject to many major hits. Later, the rock spent at least one episode under water rich with carbon dioxide. As the water seeped through, tiny pellets of carbonate formed along the rock's fissures. There the rock stayed as Mars dried up.

About 15 million years ago, the rock became an astronaut. A second impact thrust the rock into space and orbit around the Sun. It probably experienced several near-misses with Earth. Finally 13 000 years ago, the rock fell onto the Antarctic ice. Buried there all this time, it was slowly pushed toward the Alan Hills, where the ice recently melted enough to reveal it. Later studies showed that the rock was full of globules of calcium and magnesium carbonate. Around these globules, a NASA team led by David S. McKay and Everett K. Gibson used a scanning electron microscope to find strange groups of long, hotdog-shaped forms in the globules or near them. Could these be microfossils? Although they do resemble ancient Earth microfossils, they are about 100 times smaller.

Although the appearance of the fossil-like structures raises hope that the science team found life, they are probably the evidence chain's weakest link. The second line of evidence involves tiny grains of magnetite and iron sulfide. It is unusual to see these compounds existing together, and very unusual to see them together with carbonate globules. These grains are evidence for life because they are typical of the decay products of some Earthly bacteria. Not everyone agrees: others suggest that a volcanic eruption, and not any living thing, could have produced the magnetite.

Another line of evidence is the organic molecules called PAHs, an acronym for polycyclic aromatic hydrocarbons. PAHs are also byproducts of decaying organic matter. Their presence in the rock from Mars is not necessarily strong evidence for life, since these molecules might have existed in the ancient nebula that condensed to form the Solar System, and they have been found in other meteorites. So why should these particular PAHs be so indicative of life? Or, the PAHs could be simple contamination from the Antarctic ice the rock was embedded in for thirteen thousand years.

If the biological answer is correct, the temperatures in which the carbonates formed could not have been higher than 300 degrees Fahrenheit. If the temperature was much hotter, then a non-biological explanation must be found since life cannot start or exist at that temperature. Some scientists suggest that the temperature might have been as high as 1100 °C. But in March 1997, several months after the original announcement, scientists at the University of Wisconsin presented evidence that the carbonate grains did not form rapidly at high temperatures, but instead molded slowly at temperatures low enough for a biological process to occur.

Each of the three lines of evidence – shapes that look like microfossils, iron-rich grains, and PAHs – is not sufficient to conclude that the Mars rock contains evidence of life. But David McKay maintains that that all three lines of evidence taken together still point to the possibility of ancient life on Mars.

I suspect that, at the end of the day, McKay's team will be shown to be correct.

But whether or not this particular meteorite has the required evidence for life on Mars is not the point. This visitor from space has inspired scientists the world over to take more seriously the possibility that Mars might have had microscopic life in the past, and that there may even be life there now.

The Gibeon meteorite

One of the most famous meteorites is the Gibeon. It is a gigantic, 4 billion-year-old Iron Octahedrite that was "officially" found in South West Africa (now Namibia) in 1836 by Captain J. Alexander, although locals knew of this iron-rich mass long before that year. The total amount of Gibeon that fell from the sky might have exceeded 150 tons. It fell in prehistoric times, probably exploding in the atmosphere and then falling over an area so large that it offers a seemingly inexhaustible supply of samples. My wedding ring is made from a Gibeon specimen (see Figure 4.2). It is a fine octahedrite with fine scratches, called Widmanstatten lines, throughout its body. These lines are a cross-etching that add to the beauty of the rock. They result from the glacially slow cooling – about a degree per million years – of two minerals, kamacite and taenite. Washing with a weak acid makes the lines more visible.

The Tagish Lake meteorite

On January 8, 2000, a bright fireball lit the night sky in northwest Canada, exploding with a force of about a quarter that of the Hiroshima bomb. What was left of the object – some 500 small fragments – were found in and around Tagish Lake. Peter Brown, a meteoriticist from Canada's University of Western Ontario and a leader of the Tagish Lake Meteorite Recovery Operation, discovered that this object is one of the Solar System's oldest specimens, dating back 4.5 billion years.

Figure 4.2 A Gibeon meteorite wedding ring atop a Mexican fencepost cactus plant. Photo by David Levy.

How to identify a meteorite

Of all the meteors you see, the only ones you would actually be able to touch would be the ones that land, the meteorites. If you see an interesting rock that could be a meteorite, see if it passes the following tests. (In italics I offer what to look out for if you don't want your meteorite to be a meteorwrong.)

(1) Is the rock different – noticeably different in appearance – from other rocks in the area? A stony meteorite could resemble a typical dark volcanic rock, but, as a rule, a meteorite is different in appearance from surrounding rocks. *But – it could just be different – a rock moved to a new location by glaciation or some other process.*

(2) Is it heavy for its size? Most meteorites are denser and heavier than Earth-bound rocks that are the same size. Their high content of iron and nickel account for the increased density. Although this is not true for all meteorites, it is for most. *But it could be artificial iron, or slag, from somewhere; these byproducts from manufacturing processes can be easily confused with meteorites.*

(3) Many meteorites are magnetic and will respond to the pull of a magnet. Again, the presence of iron and nickel account for this, and also again, this is not true for stony meteorites. *Besides artificial iron and slag, magnetite and hematite are also magnetic. To test whether a magnetic rock is*

a meteorite or not, use the underside (rough or unfinished side) of a ceramic tile. Take the suspect meteorite and scratch it as hard as you can on the rough side of the tile. Does it leave a dark streak? Then the sample is probably magnetite. Does it leave a reddish streak? Then it is probably hematite. Unless they are heavily weathered, most meteorites will not leave any streak at all.

(4) Most meteorites tend to be irregular in shape, not round, as a result of their race through the atmosphere.

(5) Fusion coating: a thin layer of dark glass around all or part of the rock provides good evidence that as it raced through the atmosphere, its surface melted. In the final part of its flight (after it ceases being visible as a meteor), it begins to cool and the melt solidifies to form a fusion coating. If the meteorite is from an old fall, the coating my have rusted, giving the rock a dark red color. *Many ordinary rocks can also develop a thin layer that resembles a fusion crust from long periods of weathering.*

(6) The rock may have acquired an aerodynamic shape during its flight through the atmosphere. *However, if the specimen is just a fragment of a larger fall that broke apart after most of its atmospheric flight, it could have any shape at all.*

(7) Sometimes a meteorite has "regmaglypts" on its surface – markings that look like thumbprints.

Types of meteorites

Iron meteorites also contain nickel in various amounts. Perhaps the most famous iron meteorite is Canyon Diablo, which fell in northern Arizona some 50 000 years ago. It left a crater more than a mile wide, and hit the ground with such force that it overturned whole rock layers. *Hexahedrites* (with six-sided crystals), *octahedrites* (eight-sided crystals) and *ataxites* (silicated irons) are varieties of iron meteorites.

Stony meteorites make up the vast majority of the meteorites, although they are not as obvious to find. The vast majority of stony meteorites are *chondrites*, which contain tiny spherical crystals called *chondrules*, each one about a millimeter in size. Much rarer are the *achondrites*. Their lack of chondrules is an indication that they have suffered some catastrophic event on the worlds they came from, like a volcanic eruption or an impact from space. Meteorites from the Moon and Mars are examples, especially Mars, whose meteorite ALH 84001 may be the most famous meteorite ever found.

There are also *stony-iron* meteorites, a rare variety that includes the beautiful *pallasite*, an especially beautiful rock. Pallasites are named for Peter Pallas, who found the first large one in 1776. The large olivine crystals, which can be almost half an inch in diameter, can be gorgeous. The reason they are so rare is that they are fragments of the boundary between the mantle and the iron core of an earlier, much larger world that broke apart as a result of a collision.

From the giant meteorite fall of Gibeon to the microscopic particles of the Gegenschein, the smallest members of the Solar System are an important part of its character. In fact, as you see a meteor fall from the sky, you are actually witnessing, in a small way, the evolution of our solar neighborhood.

5

Observing meteors

When the moon is on the wave,
*   And the glow-worm in the grass,*
And the meteor on the grave,
*   And the wisp on the morass;*
When the falling stars are shooting,
And the answer'd owls are hooting,
And the silent leaves are still
In the shadow of the hill
Shall my soul be upon thine,
With a power and with a sign.[1]

Parts of this chapter have been updated from S. J. Edberg and D. H. Levy, *Observing Comets, Asteroids, Meteors, and the Zodiacal Light* (Cambridge: Cambridge University Press, 1994).

Visual meteor observing

In a book designed to get you excited about observing meteors, there is really no better advice than this: go outside, sit comfortably, and enjoy watching them. All the rest is detail. For me, there is no more relaxing type of observing than sitting in a comfortable lawn chair or hammock, watching the sky, and watching meteors appear out of the sky. Meteor observing can be done strictly for fun, for science, or both. If reading this book gets you interested in meteors just for fun, I would be happy. If it arouses your curiosity in how to make a contribution to meteor science, let that be a bonus. See Figures 5.1 and 5.2.

[1] George Gordon, Lord Byron, Manfred (Act I, scene I) *Lord Byron: Selected Poems and Letters,* ed. William H. Marshall (Boston: Houghton Mifflin, 1968), p. 232.

(a)

Regulus

Figure 5.1 Two images of what could be the same meteor, taken more than a mile apart. Rolf Meier took the southern one (a); David Levy exposed the northern one ((b); see facing page).

Observing a meteor shower

Shower observing is simple and easy, and yet it requires patience, care, and perseverance to do well. Observing sessions can be as simple as counting meteors with a counter and nothing else, or recording, for each meteor, the time, magnitude, shower membership, and comments on anything unusual. Regular viewing of as many meteor showers as possible, throughout the year, will help you develop and maintain a high degree of skill and enthusiasm for these observations.

(b)

In an age when electronic observing is literally taking over observational astronomy, it is good to know that visual meteor observing, the traditional way of collecting useful data on meteor streams, still plays an important role. Although nights of major showers attract many observers, observers who record meteors every clear night have the best chance of discovering new streams.

Meteor watching is just as effective whether you are alone or in a group. If you watch by yourself, do not try to scan the entire sky. Choose one quadrant and concentrate on it. Don't look directly at the radiant; that area may not produce the most meteors. Choose an area about 60 degrees from the radiant, and of course in the darkest part of the sky.

Observing alone

Watching meteors alone is a personal communion with nature in our neighborhood. Where a group observing project introduces its members to a

Figure 5.2 A Quadrantid, taken on January 2/3, 1985. Photo by Rolf and Linda Meier.

shower, an observer working alone enjoys a gentle evening out on Earth's front porch, watching meteoric visitors pass by. Choose an observing site that promises to permit a relatively high number of meteors to be seen.

Solo watching has the advantage of not needing too much planning. Depending on what shower you are watching you might see 50 meteors per hour or more under an ideal sky if you are vigilant and lucky. Individual observations can be combined without effort into certain data bases such as those maintained by the Meteors Section of the Association of Lunar and Planetary Observers, the American Meteor Society, or the International Meteor Organization.

Remember that you cannot hope to cover the whole sky. Choose a part of the sky away from city lights, one that promises the greatest number of meteors. This is not necessarily the direction of the radiant of whatever shower you are trying to observe, since meteors can occur almost anywhere. A lone observer should try to watch some 60 to 90 degrees from the radiant. The exception to this is the period from July 20 to August 14, when meteors are falling from two major radiants, the Delta Aquarids and the Perseids, and the minor radiants of the Alpha Capricornids, the Kappa Cygnids, and others, at the same time. During this period, if you want the best chance to distinguish a meteor belonging to one shower from one hailing from another, you should face the south. Meteors can be recorded either onto one of the standard meteor report forms or onto a digital recorder or magnetic tape or to be transcribed later. A tape recorder allows data to be stored without removing your eyes from the sky.

Figure 5.3 Another Quadrantid from 1985. Photo by Rolf and Linda Meier.

Do not be overly ambitious with recording your data; you'll miss meteors. Consequently, it is not a good idea to plot each meteor if you are watching alone. Actually, I will admit that my favorite way of observing meteors is with a simple gate counter as my only tool. See Figure 5.3.

Group observing

Some debate has surrounded the best way of observing in groups, with conflicting suggestions on whether observers should each record their own meteors, or have a central recorder. If groups are organized the first way, then each observer would have a separate record sheet to write down the magnitude, shower membership, and any comments about each meteor seen. If the gathering employs a central recorder, all meteors are written down on one sheet, and only one master list is kept. This way, your dark adaptation is not harmed every time you look away to record a meteor. Radio time signals or a central clock can be used by all observers.

When you record a meteor, use a red flashlight so that you can see what you are writing without unduly affecting your adaptation to the dark.

Team of five

In this arrangement, one observer takes each cardinal direction, with the fifth as recorder. Observe together in hour-long periods, with reasonable time allotted for breaks; don't try to have the observers watch for too long without a rest!

The central recorder is notified each time someone sees a meteor by the traditional call of "TIME!" The recorder then assigns a number, which the observer records if he or she plots the sighting, and then asks for the magnitude, shower membership, and any comments.

Team of ten

With this many people, you have several possibilities. First, you can have all eight observers and two recorders working at once, with the sky divided into the upper and lower portions of each quadrant. This may be the best way if you plan to observe for only one or two hours. For longer sessions, you may plan to rotate assignments, so that half the team is observing and the other half resting at any time.

The third possibility is somewhat unconventional but can be interesting and fun. You may want to divide the group into two teams at locations differing by some 40 or 50 miles, with instructions to record and plot carefully the time and path of any meteor first magnitude or brighter, and later trying to determine, through trigonometric calculation, the height and distance of these meteors.

Team of fifteen

A good way of observing with this many people is to organize a set of three groups of four each for observing and one group of three for recording. A schedule should be prepared that will allow each observer one hour on, followed by a half hour off. The hour of observing would be divided into two parts, the first for observing the upper part of a quadrant, the second for the lower, thus keeping the mind alert through changing position. A typical schedule follows, each letter standing for an observer.

Sample observing schedule

Local time	Quadrant half	North	South	East	West
2000	upper	A	D	G	J
	lower	B	E	H	K
2030	upper	B	E	H	K
	lower	C	F	I	L
2100	upper	C	F	I	L
	lower	A	D	G	J
2130	upper	A	D	G	J
	lower	B	E	H	K

Sample observing schedule *(cont.)*

Local time	Quadrant half	North	South	East	West
2200	upper	B	E	H	K
	lower	C	F	I	L
2230	upper	C	F	I	L
	lower	A	D	G	J
2300	upper	A	D	G	J
	lower	B	E	H	K
… and so on.					

The advantages of such a system are easy to see. First, the observers get equal opportunities to rest; they are on for an hour and off for a half hour. Second, the switching of observers from upper to lower halves of a quadrant adds some element of randomness to the counts, although the most astute observers will still always be in the same quadrants.

Overlapping observing areas slightly is a good idea, so that faint meteors appearing at sector edges will have less chance of being ignored. Since we are interested in meteor counts per observer, meteors seen by more than one watcher should be recorded by each of them. If a group count is kept, and meteor 41 is seen by two observers, both observers' initials should be recorded.

The order of the hole of the doughnut

To add amusement and fun to a long night of observing, the Royal Astronomical Society of Canada's Montreal Center had a special citation for observers who spotted meteor 100, 200, and so on, during a meteor watch. These privileged people got to join the "Order of the Hole of the Doughnut." Over the years, quite a number of people have been inducted into this austere order, which never had a meeting or any other function to celebrate the chance accomplishments of its members. It was an "unorder" that added to the tradition and fun of meteor observing.

Some observing hints

Time of night

The only downside to meteor observing is that it is more productive after midnight! For any meteor shower, observing becomes most productive the higher the radiant is in the sky. In most cases this means after midnight, as you are now facing slow and faster moving meteors. Before midnight the meteors you see will be those traveling faster than Earth, to overtake it as it orbits the Sun. The Perseids and Geminids offer meteors all night long. Even though meteor sightings increase after midnight, their radiants are in the sky all night long and meteors should appear in strong numbers throughout the nights near maximum.

Moon phase

It is amazing how much a bright Moon can affect observing. On bright moonlit nights you could lose as much as 90% of your meteor count, making it difficult to see any meteor fainter than second magnitude. A little planning will increase the enjoyment you have during an observing session. For example, if the night of August 12, the Perseid maximum, occurs four or five nights before full Moon, then I would suggest waiting until moonset before beginning observations. The sky will be darker, and more meteors should appear.

Comfort

As in most branches of astronomical observing, meteor watching is more effective if you are comfortable. Don't stand up and strain your neck; instead use a lawn chair or hammock and be relaxed. If you are part of a group of observers where each of several people has a different area of sky, each lawn chair can be adjusted so that observers are comfortable. The chair back can be raised to observe the part of sky below 45 degrees, or lowered for observing closer to the zenith. In 1957 the International Geophysical Year began world-wide programs designed to increase our understanding of the Earth and its atmosphere. That included the observation of meteors. Dr. Peter Millman directed the program from Ottawa, Canada, and his enthusiasm led to a very active meteor program. The observers designed a set of eight enclosures, euphemistically called "coffins," into which they could crawl. These wooden structures were completely enclosed, except for a small space for the head. The observers were thus kept warm in the cold nights that Canadians must put up with, and also helped somewhat (as I experienced during a visit to Ottawa back in 1971) to keep mosquitoes out during summer.

The "coffins" were used by the National Research Council team, and also by the Meteor section of the Royal Astronomical Society of Canada's Ottawa Centre. These were serious observers: Kenneth Hewitt-White, a lifelong first-rate observer, once observed Geminids this way for an entire night from 4:30 p.m. to 7:00 the following morning. It was worth the effort: "It pushed our total for the year to 3000 meteors!" In 1992 Rolf Meier redesigned the enclosures so that they could be taken apart for transport to darker sites.

Safety and courtesy

Especially if you are alone, but even if you are with a group, it is a good idea to let others know where you will be observing and when you plan to return. If you're planning to observe on a deserted field, make sure the owner of the field is aware of your plans and that you have permission to observe there.

Meteor observing in a group can be so energizing that the group gets raucous. In the late 1960s I was in charge of a group of youngsters observing at Camp Minnowbrook on the north shore of Lake Placid, in upstate New York. I had asked each observer to call out "Time!" whenever they saw a meteor. As the numbers grew, the "Time" calls grew more frequent and louder and the conversation more lively, the camp director interrupted to complain that we were too noisy and were keeping up the other children. Not many years later, Peter Manly told me the story of his group observing meteors from a suburban back yard. They were having fun, the meteors were falling, it was the middle of the night, and the noise level grew until a disgruntled neighbor opened his upstairs window and demanded: "Do you guys know what time it is?" As the group fell silent, their short wave radio announced, "The Time – at the tone – 3 hours, 4 minutes, coordinated Universal Time – BEEP!"

Thus, a little advice to keep good neighborly relations: if you observe in a crowded neighborhood or at at campsite with other campers nearby, keep an ear out for how much noise is going on. If, on the other hand, you are observing from a quiet place out of earshot of others, relax, have fun, and enjoy the sweeping show of the sky.

Telescopic meteors and fireballs

Telescopic meteor showers are showers whose average brightness is usually four or fainter. The June Bootids is one such shower, its meteors averaging about the fifth magnitude. The Alpha Lyrids of July is another, although it also produces a few naked eye meteors. Through binoculars or a telescope some observers have reported counts of over 10 per hour.

Some showers really do offer fainter meteors that are not visible with the unaided eye. Telescopes with short focal ratios (f/4 or f/5) and low powers are best for this type of observation. On many shower nights I enjoy using such a telescope to scan the sky for comets, and I pick up several meteors as faint as ninth or tenth magnitude during my observing time.

When observing telescopic meteors, record the aperture of your telescope, and the field of view of your eyepiece. For each meteor, write down the approximate right ascension and declination (or altitude, azimuth, and time) of

Figure 5.4 A fireball passes through Orion. Photo by Tim Hunter and James McGaha.

the center of the field of view. Note also the meteor's magnitude, although I find that determining the brightness of a meteor streaking suddenly for a split second through a telescope is very difficult. In my experience, on the nights of maxima of showers like the Quadrantids and the Orionids, I see far more meteors than on non-shower nights.

Variable star observers, who observe specific areas of sky at frequent intervals, are in an ideal position to add a telescopic meteor project to their observing program.

Photographing meteors

In this age of digital cameras, photographing meteors is fairly simple. Using a camera with as wide angle a lens as possible, take a series of photographs of the same portion of the sky, one after the other. On the night of the Orionids, I used an 80 mm telephoto lens and CCD camera to take more than 100 images of the Andromeda Galaxy, and in one of those images I was fortunate enough to capture a meteor.

There are really no rules in meteor photography, except that the more wide angle pictures you take on shower nights, the better the likelihood of recording a meteor or two. See Figure 5.4.

6

Recording meteors

Now lies the Earth all Danae to the stars,
And all thy heart lies open unto me.
Now slides the silent meteor on, and leaves
A shining furrow, as thy thoughts in me.[1]

Parts of this chapter have been updated from S. J. Edberg and D. H. Levy, *Observing Comets, Asteroids, Meteors, and the Zodiacal Light* (Cambridge: Cambridge University Press, 1994).

How to record showers

The easiest way to record a meteor shower is to use a counter of some kind. I use a simple metal counter – I push it once whenever I see a meteor. The numbers range from zero – just didn't see anything – to 2403 meteors, which my wife Wendee and I saw on the night of November 18/19, 2001. Personally, I enjoy observing in this manner most of all. It requires absolutely no auxiliary equipment except the counter.

Counter

If you are observing alone and all you want is simple hourly rates, you can use a gate counter, or a golfer's stroke counter that you press every time you see a meteor. You will get hourly rates this simple way, but not the other information which is also useful. The advantage of the counter, when pressed for each shower meteor you see, is that it allows you to observe with the specific goal of determining shower rates.

[1] Tennyson, The Princess: Now Sleeps the Crimson Petal (lines 7–10), *Victorian Poetry and Poetics* eds. W.E. Houghton and G.R. Stange (Boston: Houghton Mifflin Co., 1968), p. 44.

Tape or digital recording

Because you can superimpose your voice with audio signals from radio time signal stations like the American stations WWV and WWVH, the Canadian station CHU, Australia's VNG, and others around the world, and since it does not take your dark-adapted eye away from the stars, recording is a method preferred by many observers. The only problem is that your post-observation work is increased, since you must now listen to the tape and copy all the information on to a report form, which for a long observing night can become tedious. A microphone with an on–off switch allows you to tape only when a notation is made. Make sure that your recorder has enough electrical power – batteries fail in cold weather – and if the temperature is below freezing the recorder may not work properly in any case.

Using a report form

If you want to be more ambitious and use the IGY report form, here is how to proceed. With this extra effort, you will gain information about the shower rates per unit of time, how many meteors (shower vs. non-shower) appeared, and other interesting information about the shower.

Time

If you decide to record meteors accurately, you can use a tape or digital recorder or a central "person recorder." Announce the appearance of a meteor with a one-word alert, like "TIME!" But how accurately the appearance of a meteor should be timed is a matter of hot debate, and the range of desired accuracy is large. When the International Halley Watch was accepting meteor shower reports, its required accuracy was only one hour for their hourly count statistics; they simply wanted to know the number of meteors seen in your hour of observing. Other groups want each meteor recorded to the nearest minute or second. Perhaps Canada's Visual Meteor Program of the International Geophysical Year offered a fair compromise, asking that you record times no less frequently than every ten minutes, or five during heavy showers. If you use a tape recorder, there is no reason why you cannot have each meteor recorded to the nearest second. Simply have WWV or CHU time signals recording into the machine, as well as your meteor reports.

Always record fireballs or other unusual meteors to the nearest second. It is always possible that a fireball might land somewhere, and thus an accurate description of the path, including the time of its appearance, is essential. Other unusual meteors – long lasting ones that could be satellites reentering – should be timed accurately in case other people at another site are able to confirm the event.

Figure 6.1 An end-of-January meteor streaks through the field of Clyde, my 14-inch; photo taken by David Levy with hyperstar lens and CCD.

Weather record

This important record of your observations should be noted every half hour, or more often if conditions are changing rapidly. Cloud patterns can affect the numbers of meteors you see.

Moonlight obviously affects the number of meteors you may see. It helps not only to record the time of moonrise but also to plan your sessions around it. A Perseid watch held around the first quarter Moon on August 6 would be worth a lot more than one conducted on the night of maximum a week later with a full Moon. In the Time Record section of the IGY report form give a general indication of the lunar phase.

Lightning from distant thunderstorms could be misinterpreted as faint meteor counts; therefore its presence should be reported in your weather report. By the way, if lightning is nearby and you can also hear thunder, it is not safe to continue observing even if the sky is not completely cloudy.

Wind is another factor. If it is uncomfortably strong, it could adversely affect the quality of the observations you make, and therefore should be recorded. See Figure 6.1.

Magnitude

Estimating the brightness of a meteor should be a straightforward operation. It involves comparing the brightness of a meteor with those of stars of particular magnitudes. Before starting out, the group leader might point out some typical stars of different magnitudes. The following list might help:

Brighter than 0
Sun −26.5
Full Moon −12 to −14
Venus −4.6 (maximum brightness)
Jupiter −2.5 (maximum brightness)
Sirius −1.5
Canopus −1

0
Arcturus −0.1
Vega 0.0
Capella 0.0

0.5
Archernar 0.5
Procyon 0.5

1
Spica 1.0
Altair 1.0
Deneb 1.3
Aldebaran 1.0
Pollux 1.0

1.5
Regulus 1.4

2
Alpha Persei 2.0
Beta Aurigae 2.0
Alpha Andromedae 2.1
Alpha Arietis 2.0
Alpha Ursa Majoris 2.0
Polaris 2.0
Beta Ursa Minoris 2.0
Gamma Geminorum 1.9
Gamma Leonis 2.0
Alpha Ophiuchi 2.0

2.5
Delta Leonis 2.5
Gamma Ursa Majoris 2.4
Epsilon Cygni 2.5
Alpha Cephei 2.4
Alpha Pegasi 2.5

3
Beta Trianguli 3.0
Epsilon Geminorum 3.0
Gamma Bootis 3.1
Gamma Ursa Minoris 3.0
Alpha Aquarii 2.9
Eta Pegasi 3.0

3.5
Eta Ceti 3.4
Beta Bootis 3.4
Alpha Trianguli 3.4
Epsilon Tauri 3.5
Lambda Aquilae 3.4

Shower membership

If you have a basic understanding of perspective, you should have no trouble determining the shower to which a meteor belongs. Tracing an imaginary path back to a likely radiant is an easy process which becomes second nature after a little practice. The August showers offer more of a challenge, since two major radiants and several minor ones are active at the same time. It is possible to have difficulty deciding to which of these showers a meteor belongs, especially if it streaks toward the southwest. But even then, remember that the shower meteors have particular characteristics that could reveal their membership. If in real doubt assign the meteor to a non-shower category. Also, keep in mind that sporadics can come from any direction, including that of a shower radiant. You should suspect this if the uncertain meteor is traveling much faster or slower than typical meteors from the shower.

Comments

Qualitative impressions or quantitative data on each shower can be helpful to record. Notes on color, or meteors that fragment or break apart, should be noted in the comments section. Was a particular meteor unusually fast or slow? If the meteor was bright, did it flare or break up, or leave a train of particles that lasted more than a quarter of a minute? Positive answers to these questions should be included in the comments column of a meteor report. It helps if you discuss these special characteristics with other people on your team; this is a good reason for having a team in the first place. Chances are that bright meteors will have been seen by more than one observer, especially if you have two people in each quadrant. So discuss the meteor: what things made it appear different from the others?

Plotting

In addition to other meteor records, tracing the path of each meteor you see onto a star chart is an additional activity. The number of a particular meteor would appear inside a small circle at the start of the path, and an arrow at the other end would indicate the direction of flight.

The advantage of a plotted record of each meteor is in determining radiants. Each observer is supplied with a single chart on which all his or her meteors for the night are plotted. If the shower is heavy, change forms after a specific time and note carefully when this was done. A gnomonic projection is normally used so that the plotted meteor is a straight line on the chart. (This type of map projection ensures that any line plotted on it is a great circle. The equator and meridians of longitude and right ascension are great circles; circles of latitude and declination are small circles.)

There is a disadvantage to plotting. The process takes your carefully dark-adapted eye away from the sky for 15 to 30 seconds after you see a meteor and focuses your attention onto the illuminated sky map on your lap. During this time, of course, you may miss another meteor.

Minor shower and "sporadic" nights

Meteors can be observed any night of the year, and counting them on a night not dominated by a major shower can be especially rewarding. With experience you will have little difficulty assigning meteors to one of several active radiants, determining true sporadics, or possibly uncovering evidence of an unknown shower.

Before you begin, check Chapter 19 to find out which showers may be visible on a particular night; then spend some time acquainting yourself with the positions of their radiants in the sky. On such relatively quiet nights, plotting can be a good idea, for it enables you to determine more accurately the shower membership of each meteor.

Fireballs and trains

Since they can attract wide public attention, fireballs offer a useful way to introduce people to meteors. A fireball is usefully defined as any meteor of magnitude −4, the brightness of Venus at its best, or brighter. If a fireball is at least as bright as magnitude −8, it might actually survive its plunge through the atmosphere. The report form was designed by Dr. Peter Millman of Canada's National Research Council and is simple to use.

Bright meteors often leave trains, which are ionized high clouds in the path of the meteor. Trains usually last no more than a few seconds, but occasionally they can last for several minutes. It is definitely worth writing down when a meteor leaves a train. Sometimes a train is the only way to see a meteor. My own comet hunting has occasionally been interrupted by a bright flash, but when I look around, all I see is a faint, long cloud that, over the next few seconds or minutes, changes shape and fades.

Reentering satellites

There are thousands of satellites, spent booster rockets, "astronaut's gloves," and other baggage that now orbit our planet, enough to make the area just above our planet a junk yard. Chances are, you will see one of these bodies reenter the atmosphere some day. They are not that difficult to identify, or at least to suspect, since their orbits around the Earth are vastly different from the solar orbits of meteoroids. Compared with the speed of a natural meteor, a reentering satellite is slower, lasts much longer, and often fluctuates in brightness as it breaks up in the atmosphere. In 1965, on the roof of the science building at the State University of New York at Plattsburgh, I saw what I believe to be a reentering satellite. It refused to disappear as it traveled at leisure across the sky, finally winking out after about 30 seconds. Its cosmic job complete, the satellite ends its life, like a meteor, in a final burst of light.

Special considerations

In any amateur meteor watch, where the observers do not have the same eyesight, the same experience, the same motivation, or the same understanding of what is happening, some problems obviously appear. The most common are as follows.

Facing the radiant

Place the most experienced observer toward the radiant, or certainly within 90 degrees of it, so that shower versus non-shower meteors can be determined more precisely.

Different levels of experience

On a separate sheet at the front of a meteor shower report, indicate, on a scale of 1 to 5 (5 best), your opinion of the experience of each observer. A person on his or her first watch might be assigned a 1 or 2, while someone with more experience would get a 3 or 4. The experienced observer would get a 5. This way, the data would be reduced with a realistic idea of which data are most reliable. Don't assume that the observer with the highest quantity of meteors is necessarily the one with the highest quality of observation.

The recorder should watch carefully the patterns of each observer. See if meteors might have been missed, or if magnitude estimates are consistently too high or too low. See if the observer has a clear understanding of radiants, and is tracing paths properly back to them. As recorder, you can keep an eye on such things, for you will see for yourself some of the meteors the others are reporting.

The high counter

No matter which quadrant he or she is assigned to, this observer will always have the highest numbers of meteors. There are three possible reasons: one, this observer likely has more alert eyes and mind than the others, and is catching more objects; two, an observer sometimes catches fainter meteors than the others; or three, a vivid imagination is behind the high counts. This last condition might be difficult to detect. If the high counter is a beginner, the eyes may wander off into unassigned areas. Also, if the beginner's meteors are faint, some of them may not be real meteors. Usually, the highest counters are the most experienced, and it is wise to assign the two highest counters to opposite quadrants.

Reducing group data

The process of preparing your data for close scrutiny is laborious, but always instructive. You should list the results of your observing sessions by each type of information that was recorded. First, tabulate the meteors seen by each observer per hour, a process that will help you determine the capabilities and reliability of each person on your team. It is simplest to make starting and ending times on the Universal Time hour. If the observers record their data individually in the first place, then you are saved this aspect of reduction. After the meteors are listed in this way, they should be carefully totalled.

Reduction by observing position

Rather than offering any special value to science, this listing acts as a control, helping to check the visual acuity and accuracy of your observers.

Reduction by magnitude

Record the night's work graphically so that magnitudes are listed across the top of your page, and time listed down the page. This type of listing can show some unexpected turns in what you think a meteor shower will offer. Did this year's shower show more fainter meteors than last year's? When the total number increased after midnight, was the increase reflected across the board or only on the fainter meteors? Compare this with the observer analysis to see if your answers to these questions are affected by differences in observer perception.

Reduction by shower

When you plot shower against time, you have a chance to get a general picture of the relative strength of a shower. This reduction is especially useful when several showers are active at the same time, as happens in July and August. It also provides an opportunity to evaluate the understanding your observers have of meteor radiants.

Sample group reduction

This is a sample of the reduction of a major meteor observing session that was held on August 12/13, 1966. The work was done carefully by Isabel K. Williamson, a prominent meteor observer in Canada who introduced hundreds of people to the special joy of meteor gazing. In the days before home computers, this work was time-consuming, and this particular reduction took several nights of dedicated work to complete.

"On the invitation of Mr. Granger Robertson," Miss Williamson wrote in the September 1966 issue of *Skyward*, "the Center's meteor team observed the Perseids from his summer home at Ste. Margeurite, Que., on the night of the maximum, August 12/13.

"There was the usual overcast sky when we left Montreal. (As one of the team remarked, we wouldn't feel comfortable if the sky were clear when we set out on one of these jaunts.) We drove through the usual rain shower. We arrived at our destination and determinedly went about setting up the equipment, trying to ignore the heavy clouds. We went indoors for the usual briefing. At 9:45 p.m. EDT, one or two stars were visible and we decided to 'go through the motions' for the benefit of newer members of the team. Light rain was actually falling at 10 p.m. when we took up our observing positions but a few stars were still visible. The first meteor was called within the first five minutes and two more in the next five, which encouraged us to continue. Then the sky began to clear. By 11:30 p.m. there wasn't a cloud in the sky and we enjoyed perfect observing conditions right through until dawn. In six hours of observation we recorded

906 meteors, thus breaking our record for all showers except the famous Giacobini–Zinner shower of 1946. It was a fantastic night.

"During the night we saw passages of Echo I, Echo II and Pageos I but were much too busy with meteors to record the times of passage. We did sing "Happy Birthday" to Echo I which we had observed from Montgomery Center on the night it was launched exactly six years before."

7

A New Year gift: the Quadrantids

Th' Imperial Ensign, which full high advanc't
Shon like a Meteor streaming to the Wind
With Gemms and Golden lustre rich imblaz'd,
Seraphic arms and Trophies:
Shone like a meteor streaming to the wind...[1]

It was the 2006 Quadrantid shower that inspired me to write this book. It is not a famous shower, but it should be: it is one of the strongest showers of the year, offering up to 200 meteors per hour (190 in 1965) per observer at maximum, in competition with the August Perseids. (The rates quoted are called *zenithal hourly rates*, and assume the ideal circumstance of a single observer observing under a dark, moonless sky, limiting magnitude 6.5, while the radiant is at the zenith. Since the radiant is never at the zenith during dark hours, the actual numbers are a lot lower.)

Late in the evening of January 2, I took The Beagle, our little dog, for his evening walk. The sky was dark and cloudy, and the forecast was not good. Through small breaks in the cloud cover I looked for meteors coming from the north, from a point in space between the head of Draco and the top of the kite-shaped figure of Bootes. From this little spot in the sky, not far from the star 47 Bootis, would emanate one of the year's best meteor showers.

But not this year, and this is why so few people pay attention to "the Quads." There are two reasons. First, it peaks during the coldest time of year in the northern hemisphere, and the radiant is so far north that it is not easily visible at all in the southern hemisphere. The second reason is more critical: the strong numbers last only for a few hours on the night of maximum. Unless the Moon is out of the night sky, and unless the maximum occurs in the hours between

---

[1] John Milton, *Paradise Lost*, Book 1, lines 533–537.

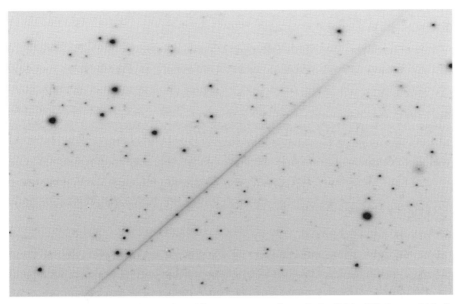

Figure 7.1 A bright Quadrantid meteor passes through the field of Clyde, my 14-inch automated Celestron telescope with hyperstar lens and CCD camera. Photo by David Levy.

midnight and dawn (when the radiant is highest in the sky) this shower will be barely detectable. Even when the circumstances are perfect, the Quadrantids (and other showers) can be poor if the Earth is not passing through a dense portion of the stream. When conditions are ideal, like they were in 1982, 1985, and 2004, displays occur under ideal conditions and are superb.

On that night at the start of 2006, the peak of the shower would take place during daylight hours over North America, so I didn't anticipate seeing anything. Our walk over, The Beagle (yes, that's really his name) and I joined Wendee for our night's sleep. But I woke up around 4 a.m. Lazily I peered out the window; the large patches of clearing I saw in the dynamic weather system moving over us, snapped me to a better state of wakefulness. Over the next 90 minutes I watched the sky as 15 meteors fell through the broken clouds, all but three Quadrantids. See Figure 7.1.

The swan song of the night was a bright Quadrantid fireball, as bright as Venus, which came out of the east and headed toward the south. It was the brightest Quadrantid I had ever seen; this shower does not produce fireballs. This was a bolide, an exploding meteor that left a shower of sparks. My total count for the night: 15 meteors, plus several more recorded by Obadiah, a telescope that was automatically searching for comets at the time. As the session went on, and especially after the fireball, I realized how much I love observing meteor showers. Why not write a book about this activity, I thought, especially on the fiftieth anniversary of my first meteor?

I have had other enjoyable Quadrantid showers. On the morning of January 3, 1982, comet discoverer Rolf Meier and I observed Quadrantids from two sites separated by a mile in the hope that we could photograph meteors simultaneously from the two sites. Then, using simple trigonometry, we hoped to calculate the distance from the meteor to us. It was a rich shower, and two meteors were bright enough (about 0 magnitude) that both cameras caught them. It was easy to see their slightly different paths through the constellation of Leo. Rolf's calculations revealed a very low point of entry into the atmosphere, which indicated that the meteors might not have been identical events after all.

Perhaps my most relaxing Quadrantid visitation occurred on the morning of January 3, 2004. I had just presented a lecture at the Longmont Astronomical Society north of Denver, Colorado, and was staying with my friends Clark and "Y" Chapman at their Rancho Europa home high in the Rocky Mountains. A snowstorm had just passed through, so severe that my flight to Denver had been delayed earlier that day. But around midnight the sky began to clear, by 4 a.m. the Moon was setting, and the sky was a dark blue so deep that the stars looked close enough to touch. With the passage of the front, the air at 8000 feet was bitterly cold, and I prepared to put on my heavy coat to face 90 minutes of freezing observing.

That was not what happened. In bathing suits, we sat comfortably in the Europa hot tub, looked up at the sky and saw 48 meteors. Most were lovely Quadrantids, all moving at a moderate speed, slower than the fast moving Leonids. The only thing between us and the dark sky was the water vapor, and that didn't prevent us from enjoying the show one bit. I kept my right hand dry and out of the water, so I could press my meteor counter each time one of us saw a meteor. I would recommend hot tub meteor observing any time, except to point out that it can be dangerous to spend more than 20 minutes in it. We lasted longer by lowering the temperature of the tub somewhat.

The shower's name

The Quadrantids offer us more than meteors; they promise us a bit of history also. It is the only shower of the year whose name is based on a constellation that no longer exists. It is an old and now abandoned constellation called Quadrans Muralis – the mural quadrant. The constellation is bordered by the western foot of Hercules, the eastern hand of Bootes, and Draco, and its area is now occupied by these constellations. Jérôme Lalande, a French Jesuit astronomer, created it in 1795 to commemorate the device he used to measure the positions of more than 47 000 stars. During a good part of this work, he worked with amateur astronomer Nicole-Reine Lepaute, one of the earliest known women stargazers. Lalande's team was most famous for its successful prediction of Comet Halley's first anticipated return in 1758, a task that he completed while Director of the Paris Observatory.

Lalande's constellation "lasted" for about fifty years. Its death knell was sounded by Friedrich W. A. Argelander, whose catalog and atlas of the stars, the *Bonner Durchmusterung*, left this constellation out. It would have been Lalande's only constellation. However, it did survive in a sense, thanks to the Quadrantid meteor shower.

Observing hints

As rich as they are, the Quadrantids are difficult because their intensity reaches such a sharp peak. For observers in the northern hemisphere, the radiant stays low in the sky, and southern hemisphere observers rarely see any activity at all.

Despite the narrow peak, rare Quadrantids can appear as early as the last few days of December and as late as the end of the first week in January. If you are observing alone, try facing a position about 90 degrees from the radiant. The southeast and northwest are preferred directions, but meteors can appear anywhere in the sky.

The parent object

On March 6, 2003, the Lowell Observatory Near-Earth Object Survey (LONEOS) discovered a distant object unromantically called 2003 EH1. Team member Peter Jenniskens compared the orbit of this asteroid with the orbit of the stream and found that the two were related enough to show that the object, which once was a comet, was indeed the parent object of the Quadrantids, and that it moves in their orbit.

According to Jenniskens' research, the shower is astronomically very young, perhaps 500 years. About that long ago, late in 1490 a comet was recorded by observers in China, Japan, and Korea. It was visible into the early weeks of 1491. No doubt it was seen all over the world, though no other surviving records have been found. In 1979, Ichiro Hasegawa suggested that the Comet of 1490 could be related to the Quadrantids. It is possible that the comet broke apart sometime before 1490, releasing the dust that we now see as the Quadrantid meteors, but so far it is hard to place 2003 EH1 near its perihelion, or closest point in its orbit to the Sun, in 1490. That doesn't preclude the comet and 2003 EH1 being the same object, however. With repeated influences by the planets, especially Jupiter, very small changes in the orbit can translate to very big and hard-to-demonstrate changes when spread out over half a millennium. A similar problem has been encountered trying to bring the orbit of Periodic Comet Levy 1 back that long; its passage in 1991 might be a return of the comet of 1499.

All this history brings us back to the predawn hours of January 3, when, if time, the phase of the Moon, and the orbits of Earth and the shower coincide, you might be treated to one of the year's biggest meteor performances.

8

The Lyrids – an April shower

Meteors make their fickle vagaries.[1]

After such a long interval – almost four months – since the last major shower, I never want to miss the Lyrids. As the night of maximum came along in 2006, I spent night after night observing the sky. Even though the darkness of the sky remained undimished by moonlight, the patterns of stars were mostly undisturbed by meteor trails. There were a few, and some were bright, but this year's shower was somewhat disappointing.

After the Quadrantids, we look forward to the Lyrids opening its curtain on the sky. These meteors can reach a theoretical rate of about 15 per hour per observer on the night of maximum, April 20/21, but I have never seen that many. In 1803 observers in eastern North America fared differently: they observed some *seven hundred* meteors in one hour. In 1982, some people noted activity reaching about 80 per hour, a dramatic increase characterized by fainter than usual meteors.

Even so, the 2006 display exceeded my experience from the year before. In 2005 I saw but one meteor. A single, solitary flash in the night was my sole experience with the Lyrids. But it was a wonderful meteor. Brighter than Vega, it sped through a few degrees of sky to the north of the radiant and left a short trail. It was greenish-white. And that was the Lyrids for 2005.

Whether the Lyrid experience is rich or miserable, the shower means a lot to me because of the influence it has had in my own evolution as a skywatcher. As a live-in patient at the Jewish National Home for Asthmatic Children in Denver, Colorado, I spotted my first Lyrid on April 22, 1963 (see Figure 8.1). "I had a regular day today," I wrote in my journal, "until tonight. I went out and saw a fireball. Then a big, fat, hunk of cloud came over. I saw no more meteors.

[1] John Bainbridge, 1619.

Figure 8.1 The Childen's Asthma Research Institute and Hospital in Denver, where the author experienced his first Lyrid meteor watch. This picture was taken in 1963, and shows Nathaniel Levy, David's father, on the left.

Tomorrow, if clear, I'll try again." The next night was again cloudy, and in my entry for April 25 I wrote "failed to see any other meteors even though I stayed outdoors for several hours." "I officially considered this year's Lyrid shower," I wrote in my diary, "to be the most disappointing failure I have ever had." I have had some finer failures since then, but what caused that sad remark? It was a cloud that happened to cover the sky throughout the session.

When my next big Lyrid adventure took place on Friday night, April 23, 1976, it was the total opposite of a "disappointing failure." It changed my life! Our group watched from the football field beside the old observatory that served as the long-time home of the Royal Astronomical Society of Canada's Montreal Center. I was at a crossroads as to what to do with my career: to continue my education in English Literature (I had a BA from Acadia University), or to try to pursue a career in astronomy. The Lyrids taught me to do both.

The Big Dipper was high overhead. As we watched meteors coming from the radiant just south of Vega, then rising in the northeast, I thought of my friend Ken Hewitt-White. Years earlier he had written a guide to basic summertime observing called *A Midsummer Night's Dream*. I loved that poetic title, and late on this particular night with the summer stars rising in the east, I thought of my talented friend and his way of placing both art and science into the sky.

As the meteors continued to fall, one at a time, I also thought of how many people had watched these same stars, and this same meteor shower, over the centuries. The oldest surviving Lyrid reports date apparently to March 16, 687 BCE.[2] Reports have been found from 1095, 1096, and 1122. The next report was the "big one" from 1803. There was another report from 1835, the year of Halley's Comet, and by that time interest in meteors was increasing thanks to the great Leonid meteor storm in 1833. After the 1835 shower, astronomer Dominique Franois Jean Arago suggested that the days around April 22 were worth keeping in mind, and in sight, for meteors. In 1839, Edward Herrick, of New Haven, Connecticut, carried out a series of observations of the shower.

By the shower of April 1976, the Lyrids had quite a pedigree, though their rates were not high. It was pleasant to look to the stars, and picture myself side-by-side with all those who had observed before, most of whom I had never heard of. On that April night all the old Lyrid nights became one, and I had the sense that our group was joined by all the Lyrid observers through history. I even wondered if the survivors of the *Titanic*, freezing on their lifeboats and on their rafts in the Atlantic on April 14, 1912, might have seen an early Lyrid or two under their heartrending moonless sky. Whether in tragedy or joy, or just plain observing, I felt joined to others who had seen an April sky.

That night planted an idea in my mind, and the very next day I decided to write my master's thesis at Queen's University on a subject that would somehow relate astronomy with literature. *The Starlight Night: Hopkins and Astronomy*, about the poetry of a nineteenth century English poet, was the result.

---

[2] Johann Galle, *Astronomische Nachrichten* **69** (1867), 382.

Astronomy and literature is a love I still have as I work on a Ph.D. thesis on the night sky in Shakespeare's time.

The parent comet

That April night was for meteors, but meteors come from comets. The year 1861 is famous in comet lore, though the parent comet of the Lyrid meteors was not initially a part of that fame. When amateur astronomer John Tebbutt discovered a comet from Windsor, New South Wales, on May 13, 1861, he had no idea that within six weeks it would brighten to become one of the century's greatest comets. At the end of June, the huge comet's head was in the northern sky near the bright star Capella, while its tail stretched more than half way across the heavens to the constellation of Hercules. Since he kept careful journals of all his observations, we are able to read about Tebbutt's own remarkable observation of the tail of his comet: "In the evening of June 30, I observed a peculiar whitish light throughout the sky, but more particularly along the eastern horizon. This could not have proceeded from the moon, but was probably caused by the diffused light of the comet's tail, which we are very near to just now."[3] The Earth, it turned out, went right through the outer reaches of the comet's tail on the night of June 30.

Communications were slow during Tebbutt's time, especially with regard to the Comet of 1861, whose discovery went virtually unknown in England until, in late June of that year, it suddenly appeared over the southern horizon with a first magnitude central part and a tail stretching over most of the sky. William Ellis, the observer assigned to England's Greenwich Observatory that night, saw the comet rise over the southern horizon. Anxious to observe it, he also feared that his employer, British Astronomer Royal George Airy, would fire him for not maintaining his prescribed sequence of observations. Torn between the work he had been assigned to do and the work he wanted to do, Ellis turned his telescope to the comet in secret.[4]

All the hoopla over the Great Comet has cast a sort of shadow over another visitor that appeared just two months earlier. On April 5, 1861, A.E. Thatcher of New York state discovered a seventh magnitude comet (now designated C/1861 G1) high in the northern sky, in Draco. The comet brightened, and was briefly visible without any optical aid; early in May it reached magnitude 3.5. As it rounded the Sun the comet was lost in twilight, not reappearing until the end of

---

[3] M. Proctor and A.C.D. Crommelin, *Comets: Their Nature, Origin, and Place in the Science of Astronomy* (London: Technical Press, 1937), p. 112.

[4] Joseph Ashbrook, *The Astronomical Scrapbook* (Cambridge: Sky Publishing, 1984), p. 46.

July, and then only in the southern hemisphere. In the meantime, the Great Comet of 1861 performed its awesome show in the northern sky.

Because Comet Thatcher appeared before, and then after, the Great Comet, some people confused the two. "The comet that flashed so suddenly upon our vision a week and a half ago," intoned the *Springfield Daily Republican*, "is now rapidly on the wane, and will soon be out of sight, lost among the constellations of the north. It has been in view just long enough to convince the astronomers that their knowledge is not infallible… The astronomers are not yet agreed whether this is the first appearance of this comet, or whether it has been seen before. The *New York Herald*, which always differs from everybody else, claims that it was seen in its approach to the sun by Mr. Thatcher, an astronomer of that city, on the fourth of April last, and calls it Thatcher's comet."[5]

The two comets are completely unrelated, even though, it turns out, both have periods of about 400 years. And while there is no evidence of particles from the Great Comet, the April comet reminds us of its visit every other April, with the Lyrid meteors. Five years after its appearance, the German observer Johann Gottfried Galle (who was the first observer of planet Neptune 20 years earlier) calculated that the Lyrid shower was linked to Comet Thatcher.

Observing hints

As I intoned earlier, don't expect a spectacular show for the Lyrids, although, if the sky is clear and moonlight doesn't interfere, it can provide an interesting show of several meteors per hour. The Lyrids are actually the opening act of a series of meteor showers; the Eta Aquarids follow early in May, and in July the summer shower activity ramps up. As with any night under the stars, remember that the sky you see, and the meteor events you record, are an experience you share with others who have looked to the stars over thousands of years.

[5] *Springfield Daily Republican*, July 10, 1861.

9

The Eta Aquarids

I am like a slip of comet,
Scarce worth discovery, in some corner seen
Bridging the slender difference of two stars,
Come out of space, or suddenly engender'd
By heady elements, for no man knows:
But when she sights the sun she grows and sizes
And spins her skirts out, while her central star
Shakes its cocooning mists; and so she comes
To fields of light; millions of travelling rays
Pierce her; she hangs upon the flame-cased sun,
And sucks the light as full as Gideon's fleece:
But then her tether calls her; she falls off,
And as she dwindles shreds her smock of gold
Amidst the sistering planets, till she comes
To single Saturn, last and solitary;
And then goes out into the cavernous dark.
So I go out: my little sweet is done:
I have drawn heat from this contagious sun:
To not ungentle death now forth I run.[1]

There is no better relation between the magic, mystery, and poetry of comets and meteors than with Halley's Comet, the most famous comet of all, and the two meteor showers with which it is related. The Eta Aquarids come first, in May (see Figure 9.1), and this shower is followed in October by the Orionids.

---

[1] *The Poetical Works of Gerard Manley Hopkins*, ed. Norman H. Mackenzie (Oxford: Clarendon Press, 1990), p. 40.

Figure 9.1 An Eta Aquarid meteor appears at the same times as a faint satellite crosses the field of view. Twelve-inch Schmidt camera photograph by David Levy.

At 04:18 on the morning of April 29, 2006, while searching for Comets through my 16-inch telescope, I saw Halley's Comet. Not the whole comet: the famous body lurked at 24th magnitude somewhere in the constellation of Hydra, some 30 times the distance between the Sun and Earth, about the distance of Neptune. Invisible now, it was less than even a "slip of comet."

I wasn't even hunting in Hydra; my telescope was sweeping through the southern constellation of Capricornus. And there, the field of view was suddenly lit by a speeding bullet of light followed by a trail of haze that dissipated within a couple of seconds. I moved the telescope back toward the northeast, along the track of the meteor, and it moved directly toward the star Eta Aquarii, the radiant of the Eta Aquarid meteor shower, within a few days of its maximum strength.

How do I know that this particular flash of light surrounds a particle that once belonged to Halley's comet? The answer takes us back to the great Herschel family, that dynasty of astronomers who so enriched our knowledge of astronomy for more than a century.

Alexander Stewart Herschel and the Eta Aquarids

With the increase in interest in meteors that took place after the great Leonid storm of 1833, astronomers began looking for other possible showers throughout the year. Astronomer Hubert Newton explored the dates of some old showers peaking in early May. He found evidence of showers as far back as 401 BCE, and later in 839 CE, 927, 934 and 1009. On April 30, 1870, Lt. Col. G. Tupman plotted meteors that came from a radiant in Aquarius. A year later he plotted more meteors from the same radiant. These observers set the stage for a connection which should be regarded as a major achievement of the Herschel family. The most famous member was of that family was Sir William, whose discoveries of Uranus, infrared radiation, and thousands of nebulae leaves him as one of history's most important astronomers. Herschel's brother Alexander, who is less well known, invented a "bell machine" or "zone clock" that his brother was able to use in his searches for nebulae across regions of the sky. Sir William's assistants raised the telescope to a certain altitude, then moved it successively up and down in two degree motions as the telescope moved through an area of sky. The assistants needed to know when the massive 20-foot telescope reach an upper or lower limit in its effort to point at a specific part of the sky. The bell machine automatically rang at both limits, alerting the assistants to reverse the telescope's direction.

This device was probably also used by William's son John Herschel, who continued the pioneering studies of the deep sky and extended them to the southern hemisphere by moving the 20-foot telescope to Cape Town during the 1830s. Sir John had a son, Alexander Stewart, who followed in his grandfather's footsteps. His early observations of the Quadrantids, Lyrids, Orionids, and Geminids encouraged the study of these major showers. In 1876, Alexander Herschel determined that the orbit of Halley's Comet comes close to the Earth in early May. His work grew out of a mathematical study that led to his conclusion that the orbit of Halley's comet intersected the Earth each May 4, when particles from the comet would be visible near the star Eta Aquarii – exactly where Tupman's meteor radiant was plotted from five years earlier.

So we can see pieces from Halley's Comet in early May, but not many. From my home in Arizona, Eta Aquarii rises just two hours before the onset of morning twilight, and that narrow window of observing opportunity does not improve much with changing latitude. In 1886, comet discoverer William Denning confirmed that Eta Aquarid meteors could be traced back to a radiant that was almost identical to that predicted by Alexander Herschel. In a triumph of collaboration between observation and theory, Denning's observed radiant agreed with Herschel's calculated radiant. Herschel suspected that particles from history's most famous comet should be visible during the first week of May, and Denning noted that the meteors he observed could be traced back to the radiant Herschel had predicted. Because they are on the outbound leg of the Halley orbit, these meteors arrive mainly in daylight; thus the night-time observation interval is short and occurs just before dawn.

The Springhill meteor observatory

This shower benefited from a major international collaboration in science called the International Geophysical Year. During all of 1957 and the first half of 1958, scientists around the world brought different methods of study to bear on the poorly understood problems about the behavior of the Earth and its vicinity during a time of sunspot maximum. Although the most famous manifestation of this work was the launch of *Sputnik* on October 4, 1957, and the launch of *Explorer* 89 days later, the actual scientific accomplishment was one of the first true international collaborations.

In Canada, Peter Millman of the National Research Council made the most of the opportunity to begin a worldwide program of observing events in the sky that were related to Earth, including the aurora borealis and, of course, meteors. He gathered hundreds of amateur astronomers, supplied them with carefully designed observing forms and instructions, and set them to work observing different meteor showers. The visual observations were valuable supplements to the radar detection facility he built at Springhill, near Canada's capital city of Ottawa. It began observations in 1958 and continued for a decade. The sophisticated radar could track meteor rates 24 hours a day, through all kinds of weather. For the Eta Aquarid shower the ability to observe during daylight hours was particularly important since the radiant is highest in the sky in the hours just before noon. The Springhill facility recorded as many as 500 meteors per hour at the shower's maximum.

This is one of two annual showers originating from Halley's Comet. Its meteors are identifiable from April 21 to May 25, with a broad maximum around May 4 and 5. These meteors appear as fast streaks, and the brightest leave long-lasting trains.

Observing hints

To find the radiant, draw an imaginary line between Scheat and Markab, the two western stars of the Great Square of Pegasus, and continue it southwards about the same distance. Just west of the end of that line, near the star Eta Aquarii, is the radiant. Chances of seeing the most meteors, however, increase if you look about 60–90 degrees from the radiant. From a dark site, that means concentrating on the overhead, northern, or southern parts of the sky. The center of our galaxy in Scorpius and Sagittarius is at a favored position at its best just before dawn, and seeing pieces of Halley's comet scratching the sky around the galactic center is a beautiful sight.

A final note

On May 4, 1970, meteors from the Eta Aquarid meteor shower were visible in the two predawn hours. They were visible on the morning of May 4, right around the maximum, throughout Ohio's university towns, for the moon was not interfering with the darkness of the clear night. By travelling out of town, an observer would have a dark sky ideal for watching meteors. For amateur astronomer Lauri Krull (now Kunkel), a Kent State sophomore majoring in fine and professional arts, that day might have offered such a pleasant time. But as she walked out of a building to go to her next class that sunny afternoon, people were racing about in all directions. Just two buildings away was Blanket Hill, where four young students lay dead after National Guard troops fired on them. "Everyone was confused and scared," she says about a day she'll never forget. "I got into my car and fled the campus."

Every May 4 at Kent State a bell rings, and at night Eta Aquarid meteors add their flashes. It marks a time that seems so distant now. The sky was Lauri's outlet: with a good eye for observing, she enjoyed watching atmospheric events like haloes, sundogs, the aurora, and meteors. Two months earlier she observed a partial solar eclipse with friends. "I love to turn people on to the sky," she explains, "especially by showing them Saturn for the first time and watching them say 'Wow!'"

10

The Omicron Draconids, continued

The curfew tolls the knell of parting day,
The lowing herd wind slowly o'er the lea
The plowman homeward plods his weary way,
And leaves the world to darkness, and to me.[1]

This chapter forms the second part of the meteor story that began in Chapter 1. Because the meteor I saw on July 4, 1958 *might* have been an Omicron Draconid, this most insignificant shower gets a whole chapter to itself. In the Chapter 1 portion, I told how the sighting of a single meteor might have ignited my interest in astronomy. The rest of the story, however, requires a little more background about the nature of meteors and meteor showers, and thus I withheld it until we reached the normal part of the year where this shower is visible.

Although I saw a meteor in 1956, it is only recently that I grew serious about trying to identify its nature. It would be easiest, of course, to conclude that since no major showers are active around July 4, this meteor was simply a sporadic meteor that is not related to any particular shower. But all meteors come from showers, even if the showers are so dissipated that they are no longer recognizable.

There are two minor northern streams that begin around July 4. To try to reconstruct the event, a few years ago I went back to the ruins of Twin Lake and stood near the same spot, trying to visualize the meteor's path toward Vega. One of the streams, the Alpha Lyrids, lasts from July 9 to July 20, but it features mostly faint telescopic meteors. Its naked-eye rate is about 1 to 2 per hour. Besides, my meteor sped toward Vega, not away from it.

---

[1] Thomas Gray, *Elegy Written in a Country Churchyard*, circa 1751.

The other shower is the Omicron Draconids. My meteor was a night short of the shower's traditional opening night of July 5/6, but the date is close enough. The shortness of this meteor's trail indicates to me that it was close to its radiant, and its slow speed is in line with the published velocity of 23.6 km/s.

The Omicron Draconid shower was discovered by a team of astronomers led by Allan Cook and Brian Marsden, who was then director of the International Astronomical Union's Central Bureau for Astronomical Telegrams. The discovery involved three meteors on Harvard College Observatory photographs taken between 1952 and 1954. Marsden suggested that there was weak evidence linking the shower to Comet Metcalf, C/1919 Q2. The difficulty with linking the shower to that particular comet is that it has a very long period, or at least it does now: it might have had a shorter period earlier in the Solar System's history. Marsden noted that though evidence for the comet connection is weak, "Allan Cook was certainly enthusiastic about it."[2]

The possible connection with Comet Metcalf is itself an interesting twist to our story. A Unitarian minister named Joel Metcalf discovered it during the summer of 1919, at another Vermont camp northwest of Twin Lake, the same year that Twin Lake was founded.

July 4, 2005

On the 49th anniversary of my meteor, I was out looking for a different type of impact. That was the night that a NASA deep space probe called *Deep Impact* careered into the nucleus of Comet Tempel 1, one of Ernst Wilhelm Leberecht Tempel's comet discoveries. That evening, Wendee and I hosted a small group of telescopes and people. We hoped to see the impact and photograph it. With Thom and Twila Peck, John and Liz Kalas, and Michael Terenzoni all peering through different eyepieces and telescopes, we waited for a small but fast moving copper projectile to smash into the comet's nucleus. About twenty-five minutes after impact, from her vantage point at the eyepiece of our 16-inch telescope, Wendee saw "a pinpoint of light in the center of the comet." It was as if someone in the comet had just turned on a light. It wasn't much – a 12th magnitude star near the center of an 11th magnitude fuzzy blob. But it doubled the comet's brightness in an instant.

Still we watched visually, taking 30-second images of the comet, one after the other. Suddenly, Mike was startled. "Wow! A meteor just passed right in front of the comet!" We couldn't believe how lucky we might be, to photograph a meteor

[2] Marsden to Levy, June 20, 2006.

Figure 10.1 A meteor passes directly in front of Comet Tempel 1, less than an hour after its collision with the spacecraft *Deep Impact*. Photo by David and Wendee Levy.

and the comet in the same telescope field – *if* the photographic telescope was photographing that instant and not downloading the previous image. I went over to the computer and waited. When the image read out a minute later, it had a bright streak passing directly over the comet! see Figure 10.1.

As I did with a meteor I had seen 49 years earlier, I tried to trace the meteor back – a difficult process since the visible trail on the photograph is only about a degree. Mike wasn't even sure of the direction of travel, but the telescope was pointed low in the west at the time. It is unlikely the meteor was going up from the southwest – more likely it was coming from the northeast. Tracing it back, the meteor could have been an Alpha Lyrid, or possibly an Omicron Draconid. See Figure 10.2.

Figure 10.2 A possible Omicron Draconid, recorded by David Levy using Clyde, his 14-inch Schmidt Cassegrain telescope with hyperstar and CCD system.

Relationship to Kappa Cygnids?

A look through the list of meteor showers in Chapter 19 shows increasing activity in late July and into August. As we shall see in later chapters, much of this activity will be from the Delta Aquarids and Perseids, but not all. One minor shower that shows up in August has a radiant only 12 degrees away from that of the Omicron Draconids. That shower is the Kappa Cygnids. The Kappa Cygnid complex has several peaks in August and can produce bright fireballs. The similarity of the velocity of the Omicron Draconids to that of the Kappa Cygnids, plus their proximity to the larger stream, would lend some support to the idea that they are part of the complex. But since the orbits of the two streams are quite different, the Omicron Draconids are probably unrelated to the Kappa Cygnids.[3]

[3] D.C. Jones, I.P. Williams, and V. Porubcan, The Kappa Cygnid meteoroid complex, *Monthly Notices of the Royal Astronomical Society*, June 2006, **371** (2), 684–694.

Does it add up?

So we are left with a small meteor shower, not even recognized by all who study meteors. It produces a few meteors now and then that may or may not be related to a comet discovered from Vermont. I cannot be certain, after 50 years, that my meteor was actually an Omicron Draconid. My 1956 observation was only two years after the meteors appeared on the Harvard plates, and well before their discovery in 1973. That would open the possibility that that at age 8, I made the first visual observation of an Omicron Draconid. The comet with which this shower is associated was discovered from Vermont, and the first visual detection of a meteor was also from Vermont. This story has a lot of "ifs" and "maybes", and the whole scenario might be fictitious. However, I did see a meteor, and there is a meteor shower at that time, and Metcalf did discover a comet. Good novels have arisen from less than that!

11

The Delta Aquarids

Watching the sky I see shooting stars, blue-green and vivid, course across the narrow band of sky between the canyon walls.[1]

The Delta Aquarids go back a long way for me. During the summer of 1963 I was a patient at the Jewish National Home for Asthmatic Children. Although visits home were not generally permitted, I was allowed to return home briefly to see the July 20, 1963 total eclipse of the Sun. While in Montreal I took advantage of the clear, hazy sky to enjoy my first Delta Aquarid observing session on July 26. In a little over two hours of observation I saw three meteors from the light polluted sky of a big metropolitan area. The next year at the Adirondack Science Camp, our team observed 13 meteors in about six hours of observation. Our night was a casual one. The count was so low because we were doing other forms of observing at the same time.

July and August are especially good months for meteor observing. As you can see in Chapter 19, several minor showers compete for attention. While the chances that, while outdoors on a July evening, you might see a meteor from a particular minor shower are low, the chances that you'll see one from *any* shower are substantial. Toward the end of July, two major showers awaken at almost the same time. The Delta Aquarids produce some bright meteors as they peak on July 28, and the Perseids begin their slow ramp upwards to their maximum on August 11/12. Separated by about two weeks, these two showers virtually guarantee that there will be good meteor activity every summer; if the full Moon interferes with one display, the other should be near new Moon. Moreover, the night of a Delta Aquarid maximum can be quite beautiful. The numbers of meteors coming from the radiant in Aquarius is supplemented by

---

[1] Edward Abbey, *Desert Solitaire: A Season in the Wilderness* (New York: Simon and Schuster, 1968), p. 163.

Figure 11.1 The Adirondack Science Camp, site of several enjoyable Delta Aquarid watches in 1964, 1965, and 1966. The camp is now the site of our Adirondack Astronomy Retreat.

meteors from three other radiants in Aquarius, and that number is almost doubled by early Perseids, so that if the sky is dark, clear, and moonless, hundreds of meteors can be seen in a single night.

The 1965 Delta Aquarids

During the summer of 1965, Glenn Myers of SUNY Plattsburgh's Physics Department and I organized a dual station meteor watch as an educational experiment for the children at the Adirondack Science Camp. See Figure 11.1. We planned to set up two observing teams, and record the time and position in the sky of each meteor seen, especially the bright ones. By comparing the two positions of any meteor seen from both locations, it was possible to calculate the altitude of the meteor. It was an ingenious exercise for the young scientists at the camp, who would use their own observations to learn something about meteors, and who would also learn to work together as a scientific team.

We set the observing date for July 27, 1965, the night of maximum. That morning, the camp had fortuitously arranged a tour of the Air Force Base Weather station. I managed to look at their weather maps and forecast, which predicted a good chance for thunderstorms that evening, and clearing later. "The trip was quite interesting," I wrote. "My plan for the afternoon (once back at the camp) was planning for the shower. Final site selection – last minute arrangements had to be completed. I took a breather and at 4:30 went rowing.

"After dinner I saw the Plattsburgh group off, and led a formal Adirondack Science Camp team meeting in the dining hall. Then the wrath of Thor was let out on us – one hour of hard rain. We set up the observing table anyway (presumably after the rain stopped). At 9:30, in spite of clouds, Glenn Myer called from Plattsburgh – go ahead. So I, with the help of my friend Steve Ashe, went around to the cabins to round the people up. At 11:00 we finally started. At 12:05 we stopped for rain – 1/2 hour. From then on the sky grew clearer gradually. Despite its thunderous start, we had a great night, with 250 meteors seen."

I couldn't wait to find out what happened. The next morning, "… after a late breakfast, spoke with Professor Myer on the phone – comparing notes. The shower observations were apparrently successful! We identified a number of meteors that were seen from both locations. In my first demonstration of the value of trigonometry, I learned how a meteor's height, and the length of its trail, are calculated."

The campers enjoyed the experience so much that we decided to repeat the observing session. We had hoped to observe on August 1, but two days of clouds prevented us from setting up until Thursday, August 3. By that time the Moon was at quarter phase, which meant that it set half way through the night, giving us a dark sky for the better half of the display.

"The clouds, after two days presence, started to leave late this afternoon, and I went ahead with plans for a meteor watch. We started observing at 21:45 and took breaks every now and then. From the start the display was good, but after about 1 a.m. (when the Moon set) they really started coming. Total count – 564 meteors."

The next year the Adirondack Science Camp did not have the opportunity to repeat the two-station experiment. But we did assemble the campers for two more enjoyable nights with the Delta Aquarids. By this time I was not a camper but a staff member, spending that summer at my first job, as an astronomy instructor. On our first night, July 22, the thin crescent Moon did not interfere and we logged 695 meteors. Four nights later we tried again. Despite the moonlight, the group recorded 782 meteors.

We tried the Delta Aquarids again in 1968, on the clear night of July 30. Like the earlier sessions, this night was used primarily as a teaching device to show young observers that meteor observing can be fun. I wanted to show them how to observe meteors, how to identify constellations in the night sky, and how to note the changing aspect of the sky as the Earth rotates eastward and different constellations rise as others set. We saw a total of 506 meteors that night.

The Delta Aquarids

Finding out who first detected a particular radiant in July is difficult, because there are actually two Delta Aquarid radiants and two more Iota Aquarid radiants. The major stream is just north of Fomalhaut, and just north of it is the northern Delta Aquarid radiant, a more minor shower. There are also the northern and southern branches of the Iota Aquarids, and the Alpha Capricornids. The radiants for all these showers are close together, which makes it very difficult for a beginner to determine which meteor comes from which shower. There are two helpful hints, however. One, by far the majority of meteors come from the southern Delta Aquarid radiant, and two, the other radiants peak on different days.

The main southern Delta Aquarid radiant has been observed pretty regularly since 1870, while the other showers escaped notice until the 1950s. Note their different characteristics in the table below.

Shower	Maximum	Velocity
Southern Delta Aquarids	July 29	41.4
Alpha Capricornids	July 30	22.8
Alpha Piscid Australids	July 30	slow
Southern Iota Aquarids	August 5	33.8
Northern Delta Aquarids	August 12	42.3
Northern Iota Aquarids	August 20	31.2
Perseids	August 12	59.4

The first three showers vie for attention around the end of July. But an Alpha Capricornid or Piscid Australid, moving at about half the velocity of the Delta Aquarids, should be easily noticed as different. In August, the other showers compete with the Perseids whose radiant, of course, is very far from the others, and Perseids move much faster. My suggestion is to enjoy the meteors of late July. Almost all of them will be Delta Aquarids and early Perseids, but there might be a few from other radiants. All these showers help make summer observing special.

12

Tears of St. Lawrence: Perseid trails and trials

It's a Colorado Rocky Mountain high
I've seen it rainin' fire in the sky
Friends around the campfire and everybody's high
Rocky Mountain High.[1]

Marathon, NY (see Figure 12.1), on the night of July 15, 1862, was a sleepy town under an incredibly dark sky. Located in Cortland county in the northwestern part of New York, the rolling hills offered peace and quiet, and starry nights. The area has grown much since that night: seven years later a major educational institution was founded in nearby Cortland. Now called the State University of New York at Cortland, the University does much to enrich an area already famous for academic pursuits and scholarship. Meanwhile, Marathon, at some distance from the county's center, still offers a sky almost as dark as it was that July night 150 years ago, when Lewis Swift turned his 4 1/2-inch diameter refractor toward the constellation of Camelopardalis, the giraffe, and discovered a comet.

On the night of the discovery, Swift thought he had picked up the already known Comet Schmidt, which had passed close to the Earth on July 4 of that year and which was rapidly fading. Three nights later Horace P. Tuttle found the same comet, from the small balcony attached to the refractor dome at Harvard College Observatory, just as he was preparing to join the Union army fighting the Civil War. Finally realizing that his comet was a new one, Swift wrote to the Dudley Observatory to report it. In England it was called Rosa's comet, for the observer who first saw it from Rome on July 25.[2]

---

[1] John Denver, *Rocky Mountain High* (New York: RCA records, 1972).
[2] S.K. Vsekhsvyatskii, *Physical Characteristics of Comets* (Jerusalem: Israel Program for Scientific Translations, 1964), p. 215; trans. *Illustrated London News*, Aug. 16, 1862, p. 179.

Figure 12.1 The Marathon train station, near the discovery site of Comet Swift–Tuttle, the parent of the Perseid meteors, in 1862. Photo by David Levy.

The 1862 appearance was marvelous because the comet passed fairly close to the Earth. The comet was as bright as the North Star and easily visible to the unaided eye. Four years later Giovanni Schiaparelli (see Chapter 2) pointed out that comets and meteors are related: his first example was the orbit of the Perseid meteor shower. It is analogous to that of Swift–Tuttle, which means that the meteors we see each August from the radiant in Perseus are residue from that comet.

What of the shower that was spawned by this comet? It is such a marvelous spectacle that historically its meteors have been called "The Tears of St. Lawrence." First written a century after his death, the Lawrence story holds that as archdeacon Lawrence awaited death by third century Roman authorities, he dispensed the Church's material wealth so that the Romans could not confiscate it. When commanded to bring the Church's treasure with him on August 10, he was accompanied by a crowd of sick and homeless people. These people, he said, were truly the jewels and treasures of the church.

On August 12, 1535, Jacques Cartier sailed towards the large river leading into southern Canada. "We named … 'Gulfe Saint Laurence his Baie,'" he wrote in his journal. Although he called the river to which it led "the great river of Hochelaga" (now Montreal), the name he gave to the bay was later applied to the whole river.[3] Although his journal does not record any meteors seen, I like to imagine Cartier looking skyward on the clear nights of that month of discovery of new lands and marveling at the "tears of St. Lawrence." Connecting the Perseids to St. Lawrence has become a tradition in some places; on the streets of San Lorenzo, Italy, a festival is held each August 10, where the August Perseids follow a day-long celebration.

I have seen Perseids coming in such rapid succession that counting and recording were difficult, followed by slack periods with little activity. I have seen many of these "tears" from my childhood home near the St. Lawrence River, the first in August 1962.

One hundred years after the discovery of Comet Swift–Tuttle, I conducted my own observing session for the great Perseid meteor shower. My grandparents, from whose cottage I was planning to observe, were not entirely happy with my plan to observe from the floating dock. On that lazy afternoon of August 12, we were on the dock at Jarnac, a remote place in the Gatineau Hills northwest of Montreal. The pond on whose surface our dock floated empties into a stream which flows into the Ottawa River, which in turn carries the water into the St. Lawrence River. As we watched clouds pass by we chatted about the magnitude −5 bolide that exploded in the western sky the previous evening. I had read in a book about how, as Perseus climbs higher in the sky, we would see more meteors; the shower would be strongest just before dawn. That meant, I insisted, that I was going to stay up all night.

As night fell on August 12 I began my vigil. Grandma and Grandpa sat with me on the dock for a short while, relaxing on deck chairs in the warm evening, but clouds prevented much observing at first. Grandpa had built a makeshift fence on the dock so that I wouldn't fall into the lake and drown. I saw my first meteor at 20:27, while the sky was still bright. As the night went on the clouds gradually dissipated, and with each passing hour the numbers of meteors increased. Five meteors left sparks. Several times two Perseids appeared almost simultaneously; for example at 03:13 I saw two first magnitude meteors, one overhead and the other toward the north. When dawn finally came, I closed my log of 112 meteors as the water below me continued on its way to joining the St. Lawrence River.

More than 40 years of Perseid meteors have passed since that August night. Some were truly memorable; in 1966 I was part of a team of observers that spotted 906 meteors. Bright moonlight interfered with some of these

---

[3] Jacques, Cartier, *A Shorte and Briefe Narration of the Two Navigations and Discoveries to the Northwest Parts called Newe France* (trans. London: H. Bynneman, 1580), p. 31.

nights; clouds ruined others. Others were with family and friends at Jarnac, during which we saw many meteors.

In 1972 the singer John Denver enjoyed a spectacular display, as noted in the verse at the start of this chapter. "Imagine a moonless night in the Rockies in the dead of summer," Denver wrote. "I had insisted to everybody that it was going to be a glorious display. Spectacular, in fact.

"The air was kind of hazy when we started out, but by ten p.m. it had grown clear ... I went back and lay down ... when *swoosh*, a meteor went smoking by. And from all over the campground came the awed responses: 'Do you see that?' It got bigger and bigger until the tail stretched out all the way across the sky and burned itself out. Everybody was awake, and it was raining fire in the sky."[4]

But clouds on the night of August 12, 1991 made the chances for my seeing *any* meteor seem slim. The previous night a large group of more than two thousand people had assembled on a mountaintop in southern Vermont called Breezy Hill. On this night of Perseid maximum, the annual Stellafane Telescope Makers Conference was over and the site was deserted when my friend Peter Jedicke and I drove back to the summit to try to catch some Perseids. Through holes in the clouds, we could see a few meteors, which was pretty unusual. Then one really bright meteor raced across the sky, its glow lighting up the cloud from atop like a searchlight beam. Around the world, other observers also noticed this burst of Perseid meteors. The reason for this outburst soon became apparent: the comet that bore them, Swift–Tuttle, was returning. On September 26, it made its first appearance through the 5-inch binocular lenses of Tsuruhiko Kiuchi's telescope in Japan. The comet was back.

The 1993 Perseid display attracted wide attention and news coverage. In California, the hoopla over the meteors and their comet had an unintended effect. On the night of August 10, Gene and Carolyn Shoemaker and I were unable to get through the front gate of Palomar Observatory because of a mob of people crowding the entrance. We finally managed to reach the gate, unlock it, and drive to our 18-inch telescope to begin our week-long observation program searching for comets and asteroids, and even an unsuccessful search with the telescope for a haze of particles around the Perseid radiant. Around midnight one of the observatory night assistants came by. He told me that as he was driving earlier along the road that joins the 200-inch with our 18-inch telescopes, using only parking lights, he almost ran over a young couple lying in the middle of the road, observing meteors, and both stark naked. The man ran into the bushes, leaving the woman to face the night assistant as best she

---

[4] John Denver, *Take Me Home: An Autobiography* (New York: Harmony Books, 1994), pp. 109–110.

could. On leaving the observatory at the end of the night, we had to physically lift a VW bug out of the way to get through the gate!

Swift–Tuttle's orbit

All this exhilaration stems from the fact that the Perseids were strong in 1993 because Comet Swift–Tuttle, one of a very few comets that return in periods between about 20 and 200 years, had returned. Other examples of such comets are Halley itself, which orbits in 76 years, Brorsen–Metcalf (70 years) and Levy 1991 L3 (51 years.) Predicting the return of this particular Halley-type comet was a mathematical triumph of Brian Marsden, who had successfully calculated the date of return based on his idea that a comet discovered by Kegler in 1737 was in fact the same as Swift–Tuttle.[5]

After it was recovered, more precise positions of its passage allowed Marsden to predicted the time of its next visit in 2126. He also determined that in the very far distant future, in the year 3044, the comet might pass within a million miles of Earth. Our descendants will hopefully have a beautiful comet to see as well as a stupendously strong meteor shower.

Perseid notes

Although history records that a Brussels astronomer named Quételet first recorded the presence of a shower from a radiant in Perseus, surely many others have noted the appearance of these meteors from much earlier dates going back at least two thousand years. The radiant is near Eta Persei, actually not far from the Double Cluster. The stream probably consists of filaments of dust spread out along the comet's orbit so that some years have better displays than others. With rates also affected by the phase of the Moon around August 12, the Perseids will offer different experiences from year to year. One thing is certain: with their slow ramping up toward maximum, their show actually lasts from the last week in July to the middle of August: at least three weeks. Even in the deep southern hemisphere, where the radiant doesn't really rise or get high in the sky, the Perseids can offer an interesting show when, just before dawn, a meteor or two can cross the sky as they graze the upper atmosphere. So whether we observe from a town in New Zealand just before dawn on August 12, or from the dark sky of upstate New York near where the comet was discovered, the Perseids offer a night to remember.

---

[5] B.G. Marsden, The next return of the comet of the Perseid meteors, *Astronomical Journal* **78** (1973), 654–662; see also IAU *Circulars* 5330 and 5586.

13

The August Pavonids

*With this, the night darkened and lights and more
lights began to flit about the wood, much as the
gaseous exhalations of the earth flit about the sky and
look to us like shooting stars.*[1]

The Omicron Draconds and the August Pavonids are definitely not major events. In this book, however, they are awarded separate chapters because they both have personal stories that illustrate different aspects of the adventure of meteor observing.

The August Pavonids quest began surreptitiously on the morning of June 10, 1991. Waiting for the sky to clear, and with dawn approaching, I opened the sliding roof on my observatory. As the southeastern sky began to clear slightly, I began looking for comets in the constellation of Aries. After about a minute I spotted a bright hazy spot that I quickly identified as Messier 74. This distant galaxy is one of the faintest of the Charles Messier catalog objects that he came upon, also while comet hunting, two hundred years earlier. I greeted this old friend ever so briefly, and since the night was ending I moved on. After another minute, I saw another fuzzy spot, a bit brighter than M74. Could I have stumbled on Messier 74 a second time? No: besides being brighter, its field of surrounding stars was clearly different. More important, where the galaxy M74 had moderately sharp edges, this object showed the gradual fading at the edge which is more typical of the gas and dust in a comet.

Being thoroughly familiar with the field of stars near M74, I knew at once that this different object was a comet, and I reported it to the International Astronomical Union's arcanely titled Central Bureau for Astronomical

---

[1] Miguel de Cervantes (1615), trans. John Rutherford, *Don Quixote* (New York: Penguin, 2000), p. 725.

Figure 13.1 Periodic Comet Levy 1, discovered in June 1991, the parent comet of the August Pavonid meteor shower.

Telegrams. It was confirmed by other observers and within a day was announced to the world as Comet Levy, 1991q (see Figure 13.1). In 1995, that numerical designation was changed to P/1991 L3.

This comet was travelling along the ecliptic, so I was not too surprised when the Central Bureau announced a few weeks later that my discovery was a new periodic comet that returns to the vicinity of the Solar System every half century. But what astonished me was the possibility that it may be the same as the comet of 1499. In that year Chinese and Korean observers observed a comet pass from Hercules through Draco, and the Little and Big Dippers.[2] The orbit of that comet, while somewhat uncertain, is so similar to that of my new find that it could be the same comet. When it returns around 2042, we'll know for sure. Its period of about 51 years means that Comet Levy is a Halley-type comet, defined as having a period of 20 to 200 years. These comets have a wide range of inclinations to the ecliptic and the orbit of the Earth. This one is inclined by 19 degrees. Halley itself, as we have seen, is the parent of two meteor showers, the Eta Aquarids and the Orionids; Swift–Tuttle commands the Perseids; and

---

[2] Donald K. Yeomans, *Comets: A Chronological History of Observation, Science, Myth, and Folklore* (New York: John Wiley & Sons, 1991), p. 410.

Tempel–Tuttle sires the Leonid shower. So it is not surprising to suspect that this new comet might also produce a meteor shower.

Shortly after the comet's orbit was announced, meteor scientist Peter Brown wrote in the *Journal of the International Meteor Organization* that there may be meteor activity associated with this comet. I remember my interest in Peter's note, until I read that the radiant was at a far southerly 65 degrees, unobservable from my location. I never thought to contact my southern hemisphere friends and ask them to observe that year. My guess is that the shower, if it happened at all, was not very strong; otherwise it would have been discovered in the normal course of meteor observing in the southern hemisphere at the end of August.

Cambridge University Press helps find a new meteor shower

Over the years I forgot about the possible meteor shower. Then came my observation of the Quadrantids in January of 2006, my realization the next day that meteor showers could be used as a tool of inspiration, and my decision to write a book about meteor showers. This year I have been gathering information on not just the major showers but some minor ones as well. Once again I came across Peter's warning 15 years ago about a meteor shower, associated with this comet, that would peak on August 31. This time I was more prepared. In June 2006, when Wendee and I visited the Royal Astronomical Society of New Zealand, we met John Drummond, the young and enthusiastic director of the RASNZ's comet and meteor section. In mid-August I contacted Drummond, who immediately alerted his observers to the possibility of some meteor activity. I was surprised and pleased by his initial report: "I believe that there is some minor activity. Last night was my fifth night in a row of observing; I'll probably try for the next two nights also – to get a good coverage on either side of the peak … My preliminary thoughts are that there are two to three meteors per hour." When his six nights of observing were completed, Drummond had observed ten Pavonids, all slow moving, at an average magnitude of 2.6. I wouldn't have expected the shower to be stronger; otherwise it would have been detected in earlier years.

Here at the Jarnac Observatory, I observed for two hours on the night of maximum, just in case a bright grazing meteor rose dramatically over the southern horizon, announcing the shower's presence as it scratched its way across the sky. Considering that even in the southern hemisphere the rate was low, it is not surprising I didn't see anything. But I enjoyed the search anyway. Sitting outside on a hammock, enjoying a late summer evening, might be the most enjoyable way to pursue astronomy. On that evening I saw a faint Kappa

Cygnid, a faint meteor that was probably sporadic, and an Iota Aquarid, but no August Pavonids.

Meteor radiants listed as theoretical do not often turn into actual producers of meteor activity. More often, an already known meteor radiant will be related to the orbit of a known comet. It was very satisfying to me to see this rarer example of observation of a comet turned into a theory of meteors that is subsequently turned again into the observation of a new meteor shower.

14

The Orionids

Him the Almighty Power
Hurled headlong flaming from th' ethereal sky…[1]

It was a very cloudy October 23 evening in 1964. I had arranged with some youthful members of my high school astronomy club, the Amateur Astronomers Association, to meet at St. Ignatius Park in West Montreal to observe this interesting meteor shower. To encourage our observations, Isabel Williamson from the Royal Astronomical Society of Canada planned to meet us at the park and record what we saw and did.

It was completely cloudy. Most people would have just canceled the watch, but not Miss Williamson, as she was always referred to. "Even if we just go through the motions," she explained, "the watch will have educational value." So we gathered at the park, warm clothing, deck chairs, and clouds, and began observing. There were four observers, one for each cardinal direction, and Miss Williamson acted as recorder. We sat and watched clouds and learned from the expert how to observe meteors. Miss Williamson knew what she was doing. She had organized dozens of meteor watches over the years, including one, on October 9, 1946, that netted more than 2000 meteors and the Chant medal, the highest Canadian honor given to an amateur astronomer.

During the night in the park some other kids came by and saw us sitting there in the cold on lawn chairs. They started to tease us, with good reason! But Miss Williamson stood up, walked over to them, and explained what we were trying to do. Instead of teasing us more, they actually got interested in what we were doing! She handled that incident very well.

---

[1] John Milton, *Paradise Lost and Selected Poetry and Prose*, ed. Northrop Frye (New York: Holt, Rinehart, and Winstron, 1951), pp. 21–22

Figure 14.1 An Orionid meteor passes in front of Andromeda Galaxy. Photo by David Levy using Lothar and Paula Eppstein's Minnowbrook lens, plus CCD camera; photo strengthened afterwards by Adam Block.

Thus my first Orionid meteor adventure ended. We stayed an hour, saw zero meteors, learned a lot about how to observe meteors in a group, and had a great time. See Figure 14.1.

Observing the Orionids

The Orionids of October are by far the best shower since the August Perseids. Meteors first begin to appear as early as the middle of October, but the numbers rise sharply as the peak approaches on October 20 and 21. Some late Orionids can fall all the way to the end of October. The radiant is at a point at the northwest corner of the constellation, north and west of the main figure of the hunter.

The Orionids probably date back to antiquity, but serious observations began during the nineteenth century by two observers, Edward Herrick and Alexander Herschel, the "forgotten Herschel" who did some of the groundbreaking work on several of the major meteor showers. In 1864 and 1865 he recorded accurately the meteors he saw – 14 the first year, 19 the second. These numbers are actually typical for the numbers of meteors per hour that the shower produces on good nights – 25 per hour would be a good shower. They are easily identified not only from their several radiants near Orion but also from their speed: at 66 km/s, they appear as fast streaks, faster by a hair than their sisters, the Eta Aquarids. This shower is a meteoric offspring of Comet Halley. And like the Eta Aquarids, the brightest of their family tend to leave long lasting trains. Fireballs are somewhat more common about three days after maximum.

Double maximum or long maxima

Unlike most meteor showers, the Orionids sometimes offer not one but two maxima that take place between October 21 and 24. On other years, the shower has one long, sustained maximum where the activity stays high throughout those three days. A. Hajduk of the Slovak Academy of Sciences examined the rates of the Orionid meteors from the first 70 years of the twentieth century. He noted that the radiant positions of the two maxima, when there were two maxima, did not recur with any fixed periodicity. His conclusion: changing activity is the result of filaments of dust existing within the orbit of Halley's comet.[2]

Although the Orionids and the Eta Aquardids are now generally accepted to be both from Halley's Comet, that was not always the case. The spring shower has orbital elements closer to those of the present orbit of Halley than do the Orionids. In 1983 Bruce McIntosh and Hajduk suggested that the orbits of the Orionid stream more closely resemble that of Halley perhaps as long

---

[2] These maxima are explained in greater detail in Gary Kronk, *Meteors: A Descriptive Catalog* (Hillside, NJ: Enslow, 1988), pp. 198–205. See also Kronk's website: www.serve.com/wh6ef/comets/meteors/showers/orionidhistory.html

ago as 1400 BC, long before we have any recorded sightings. They went on to point out that the stream has many trains of particles. This is why the Orionids, as well as the Eta Aquarids, show different levels of activity in different years.

The Orionid shower of 2006

This chapter began with my first Orionid watch, in 1964. It ends with my most recent one, in 2006. It was by far the richest display of particles from Halley's comet that I have ever seen. As the activity began to ramp up the night before maximum, with 12 meteors in an hour of observing, I had a feeling this would be a good year. On the "first night" of maximum – October 20/21, there was steady activity all night – 68 meteors in two hours. Many were faint, but there was one good bolide that exploded low in the west.

The next night – the night after maximum – offered an unexpected show. The rate was the same – 87 meteors in approximately three hours of observing, but it was by no means a constant. The meteors came in clumps and were unevenly distributed. There would be long periods of up to ten minutes or more without any meteors, then I would count them by twos and threes. That night began with Wendee and me looking at the dark sky and wondering at the constellations. Suddenly, not far east of Orion a bright yellow meteor appeared, heading west. It rushed across about 25 degrees of sky, then exploded, leaving a cloudy train that lasted for several seconds.

Although that was the prettiest meteor of the shower, it was not the brightest. Around 02:15 a magnitude −5 fireball appeared quite close to the radiant, and was therefore moving quite slowly for an Orionid. It brightened in steps, not smoothly, then burst in a bright explosion that left a train. Instead of lasting seconds, this train lasted for more than six minutes! Imagine viewing such a thing.

Also, during the night, I saw an early Taurid that appeared almost overhead and headed slowly all the way to the southern horizon. As it fell, it faded as if it wanted to disappear, then brightened again, then faded, until it finally vanished just above the mountains south of our site. Finally there were two early Leonids – although that shower was not due for almost another month! On the final night, I counted 27 meteors, but I was also concentrating on comet searching, so the maximum rate was quite possibly the same as it was for the previous nights.

The most surprising, and perhaps moving – part of the night took place in its last hours. I had been trying all these nights to get a picture of an Orionid meteor with a lens given to Wendee and me from the estate of Paula Eppstein, a dear friend who had passed away a few years ago. In 1969 her husband Lothar, about to travel to Lake Placid to run his summer camp for children, stood patiently in line at a camera store in New York to pick up his new camera, a Honeywell Pentax. Lothar didn't smile very much, but whenever I saw him walking around that camp with that camera, I knew he was enjoying himself and getting some precious moments of rest. After Lothar died in 1977, Paula and I retained our friendship and at our last visit, Paula finally met Wendee. Knowing of my own love of photography, in her will she kindly left to us that precious camera.

I have recently attached the 80 mm telephoto lens that came with the camera to a Meade DSI CCD system. I wanted to get a picture of an Orionid meteor. So as the night went on I kept on choosing fields and taking long sets of images hoping for a meteor. As I watched for meteors the night after the main maximum (the night I saw 87 meteors), I noticed that M31 was nearing an appropriate position to use as the picture. I thus began a set of 40 90-second images in the forlorn hope that I would be able to catch an Orionid as it flew through the atmosphere in front of the great Andromeda Galaxy. I then returned to where my visual comet search was taking place. As I moved through the sky in my search, I would often look up and almost each time I would see a meteor. But as I looked toward M31 I thought how small a space in the sky it took and how unlikely it would be that a meteor would appear there.

And then a third magnitude meteor appeared just east of the galaxy, crossed over it, and vanished! I couldn't believe it. I hurried back to the camera lens as it was finishing its 90-second exposure, and there was the meteor safely captured (see Figure 14.1). As I walked back to the comet hunting telescope, I realized how much fun this observing session had been. Had Paula and Lothar used their influence somehow to arrange that meteor, or do I owe my good fortune to the chances of small-number statistics and luck? It doesn't matter. For me, it is just another one of those happy coincidences that makes observing, *and particularly meteor observing*, so much fun for me – whether this time, or that cloudy night in 1964. You will have your own fun stories and adventures as you gain more experience. The important thing is, go out and enjoy the sky, and enjoy those wonderful events that take place when meteors scratch the sky.

15

The Taurids

And meteors fright the fixed stars of heaven…
I see thy glory like a shooting star
Fall from the base earth from the firmament![1]

My first good memory of the Taurids was in the fall of 1961, with Mrs. Beardsley on Oakland Avenue in Westmount, our Montreal suburb. Her son, Tony, went to school with me, and when Tony learned of my interest in astronomy, he invited me, and some members of our new astronomy club, to meet her. I was so impressed with her enthusiasm. I had no idea at the time that her family was very much involved in the Stellafane community, where amateur telescope making got its start in the 1920s, and how well known in astronomy she was.

On that first evening she told us about about the Orionid meteors and Halley's Comet, and how to observe them. Two weeks later, after some cloudy nights, she met with us again. Not giving up on the clouds, she now told us about the Taurid meteors. It turned out that the fall of 1961 and the winter of 1962 were particularly cloudy, especially on our club's Friday meeting nights, so we didn't get much meteor observing done. So my sharpest memory of the Taurids had to wait until the 1980s, particularly on the night of November 3, 1988. I was walking peacefully on William Carey Street near my home when suddenly a bolide appeared near the radiant point in Taurus. The meteor lasted more than 15 seconds, exploded, and left a train that lasted more than 15 minutes. Dogs barked in response to the flash of light! It was quite a show.

[1] William Shakespeare, *Richard II* 2.3.9, 19–20.

The Taurids and Comet Encke

This shower is actually a combination of several different streams, two in Taurus, two in Aries, but all related to some event in the history of Encke's Comet. The strongest shower, the Northern Taurids, peaks at about 15 per hour; however, a large proportion of these meteors are bright, so the shower actually appears bigger than it really is. Moreover, their speed, 28 kilometers per second, is the slowest of the major shower meteors, which gives you a better chance of seeing them as they cruise lazily across the sky. Many fireballs reported in the months of October, November, and December belong to this stream. Their different components all originate from Periodic Comet Encke, but in a special way; in 1950 Whipple and S. E. Hamid proposed that they were produced suddenly as a lot of material was rapidly discharged from the comet about 4700 years ago. A second ejection of material about 1500 years ago added to the streams. However, the parent of this new material appears to be a "child of Encke" – a different and unknown comet with greater aphelion distance, which had earlier separated from Encke.

Introducing Encke's Comet

The Taurids are one of the brightest showers of the year, but its parent comet, Encke, is certainly one of the most famous comets ever to have been viewed through a telescope. No chapter about the Taurids would quite be complete without alluding to this marvelous comet and its history.

The Taurids were first noticed in the first week in November 1869 by the Italian astronomer Giuseppe Zezioli. A southern branch was recorded by T. W. Blackhouse from England. But these showers were really too weak to be recognized until 1918, when the British astronomer Alphonso King suggested once again that a meteor shower was active during November. The reason that the shower was so late in being recognized is probably that the various radiants throughout Taurus are difficult to reconcile into a single stream, or that some orbital perturbation took place during the tenth century that sent the streams closer to Earth.

By the middle of the twentieth century, Fred Lawrence Whipple was producing his major papers on the relationship of comets to meteor showers. Whipple noticed that the orbits of meteors, which are very much like those of comets, were slowing down so quickly that all the meteors should have left our part of the Solar System, and we should never, or very rarely, see a meteor at all. Something was replenishing the Solar System's supply of meteors. Comets were then thought to be huge "flying sandbanks." Whipple proposed that if comets were made of sand, comets would not be generating meteors and there should be no meteors left. Instead, he proposed, comets were large conglomerates of ices and meteoric particles, now popularly known as "dirty snowballs."[2] By September 1951 Fred Whipple had published the results of his work with the orbit of Encke's Comet in two of the most important papers in the history of comet science.[3]

Through this work, Whipple and his colleagues established the connection between both Taurid streams and Comet Encke, more than a century and a half after the comet was first discovered on January 17, 1786, by France's Pierre-François Méchain. It was fifth magnitude that night, barely visible with the unaided eye. On November 7, 1795, Caroline Herschel of England reported her discovery of a comet. From his observing sites in France, Jean-Louis Pons found a comet on October 20, 1805, and he found yet another on November 26, 1818. Four separate comets they appeared to be, until the mathematical wisdom of Johann Franz Encke called foul – using accurate observations to calculate their orbits, Encke found that all four comets fitted the orbit of a single comet that would next return in 1822. Encke was right. Charles Rumker found the comet on June 2, 1822 and the comet which now bears Encke's name has by far the shortest orbital period of any comet, a brief 3.3 years. Encke is one of the few comets, like Halley and Crommelin, that is named for the person who calculated the orbit rather than the comet's discoverer. It was almost universally recognized that Encke's name belonged on this comet, except for Encke himself, who always referred to it as Pons' Comet in honor of the man who discovered it twice.

During the nineteenth century, Encke was recovered, or rediscovered, each time it rounded the Sun, now as a known object circling in space, by some very well-known observers. The French Astronomer Jean Elias Benjamin Valz picked it up at magnitude 7, on July 13, 1825. Otto Struve, the famous double star observer, recovered the comet on September 16, 1829. Johann Galle recovered it on February 8, 1842, just four years before he helped Urbain LeVerrier discover Neptune. The great George Bond of Harvard spotted the comet at its return on August 27, 1848.

---

[2] F.L. Whipple, *The Mystery of Comets* (Washington: Smithsonian Institution Press, 1985), pp. 145–147.

[3] Whipple announced his theory in two papers. A Comet Model. I. The Acceleration of Comet Encke, *Astrophysical Journal* **111** (1950), 375–394, explains how the orbit of Periodic Comet Encke, which is shrinking with each return, is interpreted if the structure of its nucleus consists of meteoric material embedded in ices which sublimate to gases. The freed material rushes out of the comet with some force which can accelerate the comet. The second paper, Physical relations for comets and meteors, *Astrophysical Journal* **113** (1951), 464–474, expands on this model.

The story of Horace Tuttle's recovery of Encke on January 23, 1875, takes us in a whole new direction. Tuttle had become famous as a discoverer of comets during the 1860s. Two of his comets, Swift–Tuttle in 1862 and Tempel–Tuttle in 1866, are the parents of the major meteor showers of the Perseids and the Leonids.

Shortly after his fourth comet discovery, in 1863 Tuttle became acting paymaster in the Union Navy.[4] In 1869, while serving as paymaster aboard the monitor ship *Guard*, he somehow left the ship's books with a mysterious loss of $8800.90 – a huge discrepancy at the time. Realizing it had never received the thousands of dollars it claimed Tuttle owed, the Navy ordered Tuttle to Washington for courtmartial. Stung by a growing embezzlement scandal, the military had just sentenced an army paymaster to life in a penitentiary, commuted to five years by President Grant. Tuttle's fear that the same thing might happen to him, however, did not stop him from using the Naval Observatory's 26-inch refractor to recover Periodic Comet Encke on January 23, 1875, three days into his courtmartial. Three weeks after the comet recovery, the court found Tuttle guilty of embezzlement. Because of doubts regarding his lawyer, and perhaps because of his comet discoveries, Tuttle's sentence was light: a dishonorable discharge from the Navy, approved by President Grant.[5]

As telescopes penetrated ever deeper during the twentieth century, actually "recovering" Comet Encke became less important as its orbit was so well known. But on its 2000 apparition, one of its faintest, there was a chance for one visual observer to pick it up low in the eastern sky. On the night of August 9/10, 2000, while comet hunting low in the east, I spotted a faint tenth magnitude comet. It turned out to be Encke! While recording my observations from that night, I shared for a time the stories of all those people who were involved with this marvelous comet and its fantastic meteor stream.

---

[4] Many of the interesting details about Horace Tuttle's life come from an unpublished study: Richard E. Schmidt, *H. P. Tuttle: Comet Seeker* (Washington, D. C.: U.S. Naval Observatory, circa 1980).

[5] The basis for this story comes from an unpublished memoir from Richard E. Schmidt, US Naval Observatory.

16

The Leonids

This show'r, blown up by tempest of the soul,
Startles mine eyes and makes me more amaz'd
Than had I seen the vaulty top of heaven
Figur'd quite o'er with burning meteors.[1]

As we read in Chapter 3, the Leonid shower has inspired sky watchers way back into the distant past. But it appears that the orbit of the stream changes, probably because of continued influence from Jupiter's gravity. For long periods the densest portion of the stream does not come close enough to the Earth, and then returns for a better strike. Thus, the 1833 and 1866 storms were marvels to behold, but the showers around 1899 were mild, and there was very little in 1933. And so the great Leonid meteor storms, where one meteor per second or more could be seen, entered the realm of history.

The storm of 1966

Thus, the Leonids joined other showers in the weak to moderate category. I recall, however, in one of the first *Sky and Telescope* magazines I ever received as a child, that the November 1963 shower was a little stronger than expected. The article reminded us of the great storms of the past, and wondered if 1966 might see a return to those levels of history.

The Leonids of 1964 and 1965 were even stronger, and so astronomers awaited the night of November 17, 1966, with considerable anticipation. That Wednesday evening was completely cloudy over most of the northeastern United States. In Chicago, a young Tim Hunter, who would in later years become a well-known amateur astronomer, went to sleep under a cloudy sky

[1] William Shakespeare, *King John*, 5.2.50–53.

and did not set his alarm, believing that the sky would stay cloudy. Tim was probably right. Over New York, a group of airborne observers took off before dawn and soared above the clouds to see a shower strengthening rapidly with bright firefalls as dawn grew brighter. Around 11 p.m., as the Leonid radiant rose over Las Cruces, New Mexico, phones starting ringing around the city – get out there and look at the sky! – the meteors are coming. The 60-year-old Clyde Tombaugh, discoverer of Pluto, went indoors to get his family. With each passing hour the rates strengthened, until just past 4 a.m. they reached an amazing 40 per second.

Uncertain of what to expect, Bob Goff, then a young telescope maker, started his car and headed up into the hills east of Los Angeles. But with each higher turn of the road the clouds above remained. Bob decided to head towards the highest elevation around, the road up Palomar Mountain. With each passing turn of the climb the clouds remained, and near the top of the mountain the clouds turned into dense fog. Good, Bob thought, maybe he might succeed in reaching the top of the clouds. But the thickness of the fog rendered that unlikely.

As he neared the summit, Bob noted the road straightening out. It was nearing dawn already, and there seemed no chance for clearing. Disappointed, he continued up the road, past the final turn, and headed towards the observatory's front gate. He stopped the car, got out, and looked up.

Just at that moment, the fog lifted for a minute. In a state of shock, Bob saw more than 200 meteors radiating from northern Leo but covering the entire sky. For the rest of the night, the breaks in the fog allowed him an experience he would never forget.

Back in Las Cruces, Clyde Tombaugh ended his observing as dawn brightened the sky, with bright meteors appearing even after sunrise. For the remainder of his life he would insist that, after that night in 1966, he'd never need to see another meteor shower again. In Chicago, Tim Hunter woke to read the front page story about meteors flooding the sky. Tim felt sick – did he sleep through the meteor storm of the century? Probably not – weather records indicate that the sky over much of the central and eastern portions of the United States – and Montreal, where I was – did not clear that night.

The 1999 return

By the time Comet Swift–Tuttle was recovered on its pass near the end of the twentieth century, astronomers like Rob McNaught and Peter Jenniskens had such a good understanding about the different strands of the Leonids that they predicted possibly the most unusual Leonid display ever. They had successfully predicted a strong display for 1998 and 1999, not much for 2000, and potentially very strong displays for 2001 and 2002. Their predictions were almost accurate to the minute, especially, as we shall see, for 2001.

As it turned out, for me the display began with a −9 Leonid in November 1997. In 1998 Wendee and I joined Stephen James O'Meara on Mauna Kea, Hawaii for a rich shower of bright meteors. In 1999, we were on the ship *Norwegian Dream* sailing through a mostly cloudy Aegean sea. At the very moment of maximum, just past 4 a.m., a small break in the clouds allowed Al Stern and me a one-minute meteor watch that happened to occur at the moment of maximum, but only in the constellation of Auriga. During that magic minute I saw five Leonids! It was lot of travel and preparation for one minute of observing, but technically, we did see the 1999 maximum of the Leonids.

What happened in the few years was unique in the annals of Leonid meteors. In November 2000, the Earth did not go through any major streams of Leonid particles. But in 2001 and 2002, the Leonids promised to be rich again. The 2002 shower would be hindered by a bright Moon, but in 2001, a waxing crescent Moon would set before Leo rose, allowing a pitch black sky background for what could be the best Leonid display of the last few years. The best sky in the best longitude for the display would be around Australia, so to that lovely land we headed. There would be two maxima that night, one when the Earth encountered particles ejected from the comet at its 1699 return, the other from particles that escaped in 1866.

Our 2001 experience took place at an old Indian Rock Art site near Alice Springs, Australia. We were in a group of about 30. That afternoon we had visited the Henbury craters, a collection of one large and 13 smaller impact craters about 70 miles south of Alice Springs. They mark the place where, some 15 000 years ago, a large meteorite had slammed into Earth. After we returned to our hotel in Alice Springs for dinner, we prepared to head out again to our site. Bringing blankets, tarps, and other supplies for a comfortable evening, we set up at the edge of the Ewaninga Rock Carving and Dry Lake. We had expected the lake to live up to its name, but the previous morning's weather was not as good as it was that evening. There had been quite a bit of rain, and so the center of the lake was not really dry! By about 1 a.m., we were all ready to begin. At precisely 1:30 a.m., a fireball began as a bluish blob of light low in the east. Coming right out of the rising Leonid radiant, it was moving slowly at first, then it seemed to accelerate as it arced dramatically right over our heads, sparkling like a fireworks display as it flew. Lasting over ten seconds, the meteor finally disappeared over the western horizon. Could it have returned to space after its graze with our atmosphere? It was the single most dramatic meteor event I had ever seen. "Everyone had time to look up," my wife Wendee said, "find the meteor, and shout. We knew we were in for a good show and just had to settle in and watch!"

After the bright fireball we had to wait only a few minutes before a second fireball lit up the southern sky, grazing the atmosphere like its predecessor. Meteors appeared steadily after that, but as the Earth approached the outer reaches of the dust trail that Comet Tempel–Tuttle released at its 1699 perihelion, rates increased rapidly. Unexpectedly, the 1699 meteors were mostly bright, at least first magnitude. For a short time around 2:45 a.m., the encounter with the 1699 dust trail produced steady meteors every few seconds.

The rates of meteors did not decrease after the 1699 encounter. As the time neared 3 a.m., the Earth began its encounter with a new family of Leonids – the dust particles released in 1866, the year the comet was discovered by Ernst Tempel, observing from Italy, and Horace Tuttle, who found it from the United States. Suddenly, around 3:15 a.m., a whole flock of fainter meteors with short trails appeared near the radiant. At one point, Wendee and I recorded nine meteors simultaneously, and there were many instances of five or more meteors coming within a second's time. From his site on a remote landing strip half way between Alice Springs and Darwin, Tom Glinos was mesmerized by the storm, even with his 30% cloud cover. "It was just awesome," he reported. "Like watching fish shooting out of a barrel. At times it moved me to tears." The 1866 dust trail provided an incredible burst of activity superimposed on an already spectacular night.

Meanwhile, the Southern Cross rose in the southeast, dragging Beta and Alpha Centauri with it. It was at this point that I decided to tell the group the story of Bart and Priscilla Bok. They had made a single star and nebula, Eta Carinae, their life's work. I hoped he was there that night, enjoying a ringside view of one of Nature's most exciting places. When I asked the group to look toward the Eta Carinae part of the Milky way, two bright meteors cruised right next to it! Eventually the full majesty of the southern polar region became visible – Alpha and Beta Centauri, the Cross, the Coal Sack, the Eta Carinae region of the Milky Way, and the Large and Small Magellanic Clouds – all forming an arc around the south pole, and all an exquisite backdrop to a storm of meteors.

So many meteors were falling that from our site some of the longer-trailed ones seemed to converge in the west in a form of "anti-radiant." Someone saw a meteor appear near the western horizon. As it disappeared behind the horizon a momentary bright glow appeared. It must have exploded just after it disappeared from our view.

The zodiacal light brightened, then dawn began, and still the rates kept up. They were still falling into bright twilight. In fact, Wendee and I counted over

Figure 16.1 Our counter recorded the record 2164 meteors that Wendee and I saw during the Leonid storm of November 18/19, 2001.

Figure 16.2 A Leonid fireball splits as it is captured in this photo by Tim B. Hunter and James McGaha.

three hundred meteors between the start of astronomical twilight and our last one just 25 minutes before sunrise. On member of our group actually saw a meteor *after* sunrise. Over that night, Wendee and I saw 2164 meteors (see Figure 16.1).

Obviously, that November night was the most memorable night of meteors Wendee and I have ever experienced. But that does not mean there haven't been others. One evening while driving near our home we witnessed a fireball brighter than the quarter moon. A ten second observing session! Another time, Wendee came out to observe Orionids with me. As soon as we started observing we saw a long, graceful, exploding fireball. Wendee went in soon after; I stayed out two more hours and never saw a meteor as brilliant as that one. All meteor nights are different and fun in their own way, from a peaceful view of a single meteor on a dark night to the magnificent meteor storm of 2001. See Figure 16.2.

17

The Geminids

My lord, do you see these meteors? do you behold these exhalations?[1]

During the autumn of 1983, planetary scientist Clark Chapman and I were observing asteroids from Kitt Peak, the U.S. National Observatory in southern Arizona. As we recorded the magnitudes of each of our asteroids during the night, Clark told me of the discovery of a three-mile wide asteroid called 1983 TB, an asteroid that seems to share the orbit of the Geminid stream. The major shower of the year, the Geminids would also be the most famous if the maximum were not during the bitter cold of winter in the northern hemisphere. But when the sky is clear, a bright, slow moving Geminid fireball crossing the sky is truly one of nature's most awesome sights. If this new object is the parent, the Geminids would be even more interesting. Moreover, would there be a chance that this object, now known as asteroid 3200 Phaethon, become the "mother of all Geminids" and collide with the Earth someday? Or could its orbit change so that, in a century or so, we won't even see the Geminids any more?

Phaethon was discovered in Draco on October 11, 1983 by John Davies and Simon Green, using IRAS (the Infrared Astronomical Satellite). The object, first known tentatively as 1983 TB until it received its permanent designation as 3200 Phaethon, was moving in an orbit so closely matched to the orbit of the Geminid stream that it was obvious that the parent had at last been found. It is one of the smallest orbits of any asteroid, having a revolution period of 1.4 years. The orbit takes Phaethon close to the Sun once each orbit. Meteor streams tend to come from comets, and this object has not shown any cometary activity – neither coma nor tail – at all since it was discovered. Most

---

[1] William Shakespeare, *Henry IV Part I*, 2.4.310–311.

scientists now believe that Phaethon was once an active comet, but that the repeated close encounters with the Sun caused it to lose its cometary material, leaving only its rocky core. The meteors we see came from that active period in its history. Others suspect that Phaethon is indeed an asteroid, and that pieces of it fall off to become meteoroids each time the comet gets close to the Sun. And it does get close to the Sun – in fact, at its closest it is about one-seventh the distance between Earth and the Sun. It is possible that, at such a close distance, small pieces of rock would leave the asteroid, adding strength to the Geminid stream.

The discovery of Phaethon was a highlight of a long story. In 1950, Miroslav Plavec, a Czech astronomer, suggested that the parent comet was once in a very different orbit – perhaps it was a big comet, like the comet of 1680, on a parabolic orbit that got very close to the Sun. Years later, comet discoverer L'ubor Kresák put forward a different theory. He suggested that the orbit of the Geminids did not get perturbed into its present Earth-crossing orbit from a different orbit, and that the parent comet should be somewhere in that orbit. In October 1983, the discovery of Phaethon proved him right.

Each year, the Earth crosses the orbit of this stream. The radiant is about a degree northwest of Castor, certainly the easiest of all the major radiants to spot. The radiant stays low in the north through most of the southern hemisphere, and although the shower is not as strong as it is up north, the Geminds are well worth watching anywhere. However, it wasn't always that way. Besides a few reports of ancient fireballs, there are virtually no reports of Geminid meteors before 1862, but the shower has since strengthened remarkably, until by the 1930s some observers reported as many as 70 per hour. That rate has steadily increased. In 2006 some observers reported zenithal hourly rates of well over 100 per hour. See Figure 17.1.

By the 10th of December each year, the Geminids are becoming a force to be reckoned with in the night sky. Because the radiant rises early, by 9 p.m. local time the numbers of meteors start to climb, and, as the hours of night pass, more and more meteors appear. In December 1987, Thom Peck (now a master optician) headed west from a Chicago suburb to try to escape some clouds and catch some Geminids. "I drove out about 15 miles to escape the light and clouds. I found a spot; the sky had cleared, and the rate was about one Geminid each minute or two. After 10 or 15 minutes, a barrage of meteors suddenly began coming down – in the next five minutes I saw 85 meteors. Then the rates suddenly returned to what they were before."

I have more pleasant memories of the Geminid shower than I can recall. Most of them took place after I moved to the warm climate of southern Arizona in 1979. Although the nights are still cold, they are not as frigid as those I

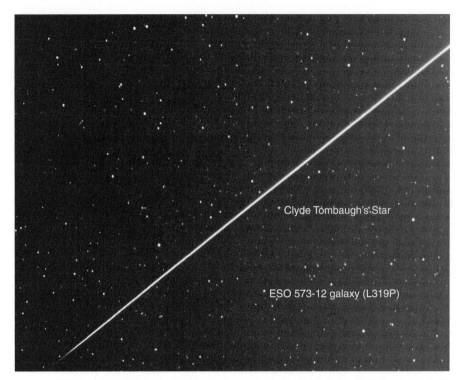

Figure 17.1 A bright Geminid meteor passes through the field of TV Corvi, a variable star that undergoes outbursts at irregular intervals. It was during my research for a biography on Clyde Tombaugh, discoverer of the planet Pluto, that I uncovered his discovery of this star in 1931. I have been watching its behavior since 1989.

remember in Montreal when, in 1961, I spent a frigid hour at Murray Park in Westmount to see 15 Geminids. These graceful, slow-moving meteors provided my introduction to the Geminids. On a good night you may see as many as two per minute. Also, they tend to cluster; a period of few or no meteors will be followed by a rich stretch of meteors that seemingly follow each other.

Like all meteor showers, the number of Geminids you see depends on the phase of the Moon. In 1999, Bob and Lisa Summerfield joined Wendee and me for a beautiful show of Geminids that lasted pretty much the entire moonless night. Our total was 910 meteors. In 2006, Wendee and I, along with astronomer Michael Terenzoni of the Flandrau Science Center, enjoyed more than two hours of observing during which we saw 109 meteors. Both nights were beautiful, made more so by the hour as Orion, the winter Milky Way, and Gemini rose higher and higher in the sky. With each hour also, more and more Geminids appeared.

Despite the northern hemisphere cold, the Geminids are worth the effort. Dress warmly and plan to spend a chilly hour or two watching the great show of this meteor shower. And let's be happy that we live in a short period of time where the Geminids are so strong. As their orbit slowly perturbs, in a century or so we may lose the Geminids, and the risk of being hit by Phaethon, altogether.

18

The Ursids

I shall fall
Like a bright exhalation in the evening,
And no man see me more.[1]

With so much attention being paid to the Geminids peaking on December 13, we often forget the last major shower of the year, the Ursids. Discovered by the famous British astronomer William Denning around 1900, the shower was considered minor until the extraordinary night of December 22, 1945. Antonin Becvar was observing at the Skalnate Pleso Observatory in Czechoslovakia, and it was not hard to notice rising numbers of meteors coming from the north. He arranged for photographs to be taken and proper counts to be made of meteors coming from a radiant near Beta Ursae Minoris, the second brightest star in the Little Dipper after Polaris. A few years later Becvar became famous as the author of several star atlases, including the *Skalnate Pleso Atlas of the Heavens*, which served for most of the second half of the twentieth century as the leading star atlas for serious astronomers. Becvar also headed a most serious comet search program from Skalnate Pleso, which resulted in the discoveries of several comets from that site.

After his rediscovery of the Ursids, observations visual, photographic, and radar came within a few years. They are now considered a respectable meteor shower, the most recent of the major showers to be detected. Activity varies from year to year. Some years the rates may be below five per hour, but some years are far more favorable.

[1] William Shakespeare, *Henry VIII*, 1.2.225–228.

Comet Tuttle

After the Perseids and the Leonids, the Ursids are the third shower to be associated with Horace Tuttle (see Chapter 12). The respective orbits of the comet and the shower are such that the heaviest Ursid activity years tend to occur six years *after* the comet's perihelion passage. Astronomically, this is very odd; the Leonids and other major showers perform best within a few months, or maybe as much as two or three years, but rarely when the comet is near its farthest point from the Sun. Since the comet has a 14 year period, this places the comet near that point – its aphelion.

In 2002 a team led by Peter Jenniskens tried to explain this anomaly. By calculating the positions of the comet's dust trails, they concluded that the meteoroids are trapped in an orbital resonance with Jupiter, but that the comet's resonance is different from that of the meteor stream. Thus it takes about 600 years to decrease perihelia (closest points to the Sun) so that they begin to intersect the Earth's orbit, and this separates the meteors from the comet by quite a distance.[2]

Comet 8P/Tuttle should actually be called Méchain-Tuttle since the famous early comet hunter Méchain discovered it on January 9, 1790. Horace Tuttle disovered the comet at a much later return, on January 5, 1858, a few months before the great comet Donati appeared.

Observing the Ursids

Because the Ursids come just one week after the Geminids, it serves as an alternative show in case the Geminid sky was cloudy or lit by a full Moon. The Ursids are rarely as strong as the Geminids, but they are enjoyable to watch. In 1986, some observers reported high zenithal hourly rates.

The battle of north and south

Twenty years later, the year 2006 was a strange one for the Ursids. On the night of the Geminid maximum on December 13, I saw several Ursids, including one zero magnitude one. That was the night I saw the "battle of the meteors." Geminids were coming from the east, and besides the bright Ursid from the north, a bright Monocerotid rushed up from the south. The Ursids were early (a week before maximum), the Monocerotids were late, and the busy

---

[2] P. Jenniskens *et al.*, Dust trails of 8P/Tuttle and the unusual outbursts of the Ursid Shower, *Icarus* **159**, 1, September 2002, 197–209.

Geminid population made for a most exciting night. With all this activity, I expected a beautiful Ursid maximum a week later.

I was disappointed. In 1.5 hours of observing, I saw only one fifth magnitude Ursid! However, there was also a single second magnitude Monocerotid to add spice to the cold night. It was an interesting experience for me, but not one that would inspire many people to take up meteor observing. It taught me that the Ursids are a most variable meteor shower. I actually saw several more Ursids on the night of the Geminid maximum than I did on the Ursid maximum. I have been more successful on other Ursid years.

I suggest that the Ursids could be a good observing project from mid-to-late December. From the maximum of the Geminids on December 13 to about Christmas night, meteors out of the north can appear. Although most of their meteors are fainter than third magnitude, some bright ones do appear each year, which makes this shower worth the effort of watching. Actually, the Ursids can add some spice to the silent, snowy nights of the holiday season.

19

A catalog of meteor showers throughout the year

I missed the meteor once …[1]

I used three different sources to list these meteor showers. Where asterisked, information is from Alan F. Cook's *A Working List of Meteor Streams.*[2] The + refers to Peter and Margaret Brown's recently completed radar study of meteor showers from 2001 to 2006. The other is Gary Kronk's *Meteor Showers: A Descriptive Catalog.*[3] The three lists are far from identical and represent studies done with different equipment (visual, photographic, and radar), and they do produce some different statistics. Moreover, Brown's meteor velocities are "out of atmosphere" velocities, while Cook's are directly measured velocities taken once the meteor has entered the atmosphere.

There are a lot of showers represented in this list, but the list does not include every shower that has been suggested or observed. However, it does offer the showers that are most likely to produce meteors that an observer can see visually. It is my hope that the printed list with its information and comments, along with the graph (Figure 19.1), will be a useful planning aid when you decide to watch for meteors on any given night of the year.

Zeta Aurigids

Maximum: New Year's Day.
Lasts from: December 11 to January 21.
Rates: 1–5 per hour.
Velocity: slow moving.

---

[1] William Shakespeare, *Henry VIII*, 5.5.50.

[2] A.F. Cook, Working List of Meteor Streams, in *Evolutionary and Physical Properties of Meteoroids* eds. C.L. Hemenway, P.M. Millman, and A.F. Cook (Washington, DC: NASA SP-319, 1973), p. 183.

[3] Gary Kronk, *Meteor Showers: A Descriptive Catalog* (Hillside, NJ: Enslow, 1988).

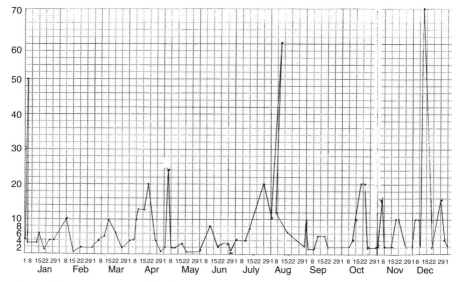

Figure 19.1 A graphical representation of meteor activity in a typical year. This figure is intended to give a "quick-look" idea of what the meteor strength might be on any particular week in a year. Moonlight, weather, and variable shower strength will make a big difference from year to year.

Comments: minor shower, fireballs rare but possible. William Denning first identified this shower, the first of the new year, early in 1886, and Alexander Hershel observed a fireball from the Zeta Aurigid radiant. The stream has a northern and a southern branch.

Quadrantids
Maximum: January 2/3.
Lasts from: December 28 to January 7.
Rates: up to 40–200 per hour.
Velocity: 41.5 km/s,* 43 km/s.[+]
Comments: meteors tend to be bluish; see Chapter 7.

Epsilon Leonids
Maximum: January 3.
Lasts from: January 1 to January 5.
Velocity: 54 km/s.[+]

Alpha Cetids[+]
Maximum: January 5.
Lasts from: January 3 to January 11.
Velocity: 46 km/s.[+]

Theta Corona Borealids[+]
Maximum: January 5.
Lasts from: January 3 to January 22.
Velocity: 40 km/s.[+]

Gamma Velids
Maximum: January 5–8.
Lasts from: January 1 to 17.
Rates: scarce.

Rho Geminids or January Cancrids
Maximum: January 8/9.
Lasts from: December 28 to January 28.
Rates: very low.
Comments: in 1963 Gerald S. Hawkins, who wrote *Stonehenge Decoded* detailing his theory that the great English monument was once used as an astronomical observatory that could predict eclipses, was one of the discoverers of this shower.

January Draconids
Maximum: January 13–16.
Lasts from: January 10 to January 24.
Rates: less than 1 per hour, scarce.

Zeta Corona Borealids
Maximum: January 14.
Lasts from: January 11 to January 24.
Velocity: 46 km/s.[+]

Beta Bootids[+]
Maximum: January 16.
Lasts from: January 5 to January 18.
Velocity: 43 km/s.[+]

Eta Craterids
Maximim: January 16/17.
Lasts from: January 11 to January 22.
Rates: 1 or so per hour.
Velocity: rapid, streak-like.
Comments: meteors tend to be faint, fourth magnitude or fainter. Denning also suspected a second radiant he called Zeta Craterids.

January Bootids

Maximum: January 16–18.

Lasts from: January 9 to January 18.

Comments: shower is of short duration. Since the stream is fairly young, it is possible that the parent comet might be found someday.

Delta Cancrids*

Maximum: January 15/16.*

Lasts from: January 13 to January 21.*

Rates: low, maybe 1 or 2 per hour.

Velocity: slow, 28 km/s.*

Comments: can be bright.

Alpha Hydrids

Maximum: January 20/21.

Lasts from: January 15 to January 30.

Rates: 2–5 per hour.

Velocity: 46 km/s.[+]

Comments: colors vary among white, yellow, and blue.

Eta Carinids

Maximum: January 21/22.

Lasts from: January 14 to January 27.

Rates: 2 per hour.

Comments: meteors are usually white.

Canids

Maximum: January 24/25.

Lasts from: January 13 to January 30.

Alpha Leonids

Maximum: January 24–31.

Lasts from: January 13 to February 13.

Rates: scarce (telescopic shower), maybe 3 per hour.

Aurigids

Maximum: February 5–10.

Lasts from: January 31 to February 23.

Rates: 2 per hour.

Velocity: slow.

Alpha Centaurids

Maximum: February 8/9.

Lasts from: February 2 to February 25.

Rates: 3 per hour.

Comments: southern hemisphere shower, visible from latitudes south of the Florida Keys.

Beta Centaurids

Maximum: February 8/9.

Lasts from: February 2 to February 25.

Rates: up to 14 per hour.

Comments: less consistent in activity than the Alpha Centaurids.

Delta Leonids

Maximum February 22/23, February 25/26.*

Lasts from: February 5 to March 19.*

Rates: 3 per hour.

Velocity: slow, 23 km/s.

Sigma Leonids

Maximum: February 25/26.

Lasts from: February 9 to March 13.

Rates: occasional.

Comments: not the same shower as the April Sigma Leonids.

Rho Leonids

Maximum: March 1–4.

Lasts from: February 13 to March 13.

Rates: very scarce.

Pi Virginids

Maximum: March 3–9.

Lasts from: February 13 to April 15.*

Rates: 2–5 per hour.

Velocity: moderate, 35 km/s.

Leonids–Ursids

Maximum: March 10/11.

Lasts from: March 18 to April 7.

Camelopardalids*

Maximum: none.

Lasts from: March 14 to April 17.*

Rates: rare.

Velocity: very very slow (6.8 km/s).

Gamma Normids
Maximum: March 16/17.
Lasts from: March 11 to March 21.
Rates: 5–9 per hour.

Delta Mensids
Maximum: March 18/19.
Lasts from: March 14 to March 21.
Rates: 1 or 2 per hour.
Comments: southern hemisphere only.

Eta Virginids
Maximum: March 18/19.
Lasts from: February 24 to March 27
Rates: 1 or 2 per hour.

Beta Leonids
Maximum: March 19–21.
Lasts from: February 14 to April 25.

Theta Virginids
Maximum: March 20/21.
Lasts from: March 10 to April 21.
Rates: 1–3 per hour.
Comments: at this time of year several minor showers emanate from Virgo. They might be part of a great complex of meteor streams, coming from different visitations of an ancient comet whose dust particles escaped during different returns.

Eta Draconids
Maximum: March 29–31.
Lasts from: March 22 to April 8.
Rates: scarce.

Delta Draconids*
Maximum: none.*
Lasts from: March 28 to April 17.
Velocity: moderate, 26.7 km/s.*

Tau Draconids
Maximum: March 31–April 2.
Lasts from: March 13 to April 17.

Kappa Serpentids[*]
Maximum: none.[*]
Lasts from: April 1 to April 7.
Velocity: moderate (45 km/s).

Delta Pavonids
Maximum: April 5/6.
Lasts from: March 21 to April 8.
Rates: 1–10 per hour.
Comments: possibly from Comet Grigg–Mellish.

April Virginids
Maximum: April 7/8.
Lasts from: April 1 to April 16.
Rates: rare.
Comments: part of the Virginid complex.

Virginids[*]
Maximum: none.[*]
Lasts from: February 3 to April 15.
Velocity: moderate (35 km/s).
Comments: some sources lump all the Virginid radiants together into a single meteor shower.

Alpha Virginids
Maximum: April 7–18.
Lasts from: March 10 to May 6
Rates: 5–10 per hour.
Velocity: slow.
Comments: part of the Virginid complex.

Gamma Virginids
Maximum: April 14/15.
Lasts from: April 5 to April 21.
Rates: 3–4 per hour.
Comments: Virginid complex.

Sigma Leonids[*]
Maxiumum: April 16/17.[*]
Lasts from: March 21 to May 13.[*]
Velocity: slow; 20 km/s.[*]
Comments: not the same shower as the February Sigma Leonids.

Librids
Maximum: April 17–18.
Lasts from: March 11 to May 5.
Rates: scarce.

April Ursids
Maximum: April 19/20.
Lasts from: March 18 to May 9.
Rates: scarce.
Comments: this shower has a long duration; meteors visible over several weeks.

Lyrids
Maximum: April 21/22.
Lasts from: April 16 to April 25.
Rates: ~20 per hour.
Velocity: 49 km/s.[+]
Comments: Comet Thatcher. See Chapter 8.

Pi Puppids
Maximum: April 23/24.
Lasts from: April 18 to April 25.
Rates: very low, except in 1972 (18–42 per hour).
Comments: P/Grigg-Skjellerup.

Mu Virginids*
Maximum: April 24/25.*
Lasts from: April 1 to May 12.
Velocity: moderate (29 km/s).*

Alpha Bootids
Maximum: April 27/28.
Lasts from: April 14 to May 12.
Velocity: a little slow (20 km/s).*
Comments: (slight suggestion of relation to 73P/Schwassmann–Wachmann 3).

Phi Bootids
Maximum: April 30/May 1.*
Lasts from: April 16 to May 12.*
Velocity: very slow (12 km/s).*

Alpha Scorpiids
Maximum: May 3.

Lasts from: April 11 to May 12.
Velocity: moderate (35 km/s).*

Eta Aquarids

Maximum: May 5/6.
Lasts from: April 21 to May 12.
Rates: 20 per hour.
Velocity: moderate (66 km/s).[+]
Comments: 1P/Halley. See Chapter 9.

May Librids

Maximum: May 6/7.
Lasts from: May 1 to May 9.
Rates: 2–6 per hour.

Eta Lyrids

Maximum: May 8–10.
Lasts from: May 3 to May 12.

Southern May Ophiuchids

Maximum: May 13–18.
Lasts from: April 21 to June 4.
Rates: scarce.

Epsilon Aquilids

Maximum: May 17/18.
Lasts from: May 4 to May 27.
Rates: telescopic.

Northern May Ophiuchids

Maximum: May 18–19.
Lasts from: April 8 to June 16.
Rates: 2–3 per hour.

Chi Scorpiids

Maximum: May 28–June 5 (June 5).*
Lasts from: May 6 to July 2 (May 27 to June 20).*
Velocity: slow.
Comments: closely related to Omega Scorpiids. (Chi is northern part.)

Omega Scorpiids

Maximum: June 3–6.
Lasts from: May 19 to July 11.

Comments: more unpredictable than the Chi Scorpiids. (Omega is southern part.)

Librids

Maximum: June 7/8, 1937.

Lasted from: June 7 to June 9, 1937.

Velocity: very slow.

Comments: identified again in 1992 (by M.A. Cervera and W.G. Elford).

Tau Herculids

Maximum: June 9/10 (June 3).*

Lasts from: May 19 to June 19 (May 19 to June 14).*

Rates: variable – highest was almost 50 per hour in 1930.

Velocity: slow.

Comments: derives from 73P/Schwassmann–Wachmann 3. Faint; average fourth magnitude. With the parent comet dividing into many pieces and passing close to the Earth in May 2006, I watched carefully for Tau Herculids on the night of June 9, 2006, and saw a single meteor. Not a great show, but it was good to see a piece of Comet SW-3 entering our atmosphere as the comet itself was visible in the sky.

Theta Ophiuchids

Maximum: June 10/11 (June 13).*

Lasts from: May 21 to June 16 (June 8 to 16).*

Rates: up to 10 per hour.

Velocity: moderate.

Comments: part of Cuno Hoffmeister's "Scorpius–Sagittarius system" that includes this region and which is active between April and July. Hoffmeister thought the Theta Ophiuchids to be at the core of this complex.

Sagittariids

Maximum: June 10/11.

Lasts from: June 10 to 16 (June 8 to 16).*

Rates: rarely seen.

Velocity: moderate–fast (40 km/s).[+]

Comments: active mostly in 1957 and 1958.*

June Lyrids

Maximum: June 15/16 (June 11–21, 1969).*

Lasts from: June 10 to June 21.

Rates: 2–3 per hour; variable.

Velocity: moderate.

Comments: sharp peak. Possible relation to Comet Mellish (C/1915 C1). Mostly 1969.*

June Aquilids
Maximum: June 16/17.
Lasts from: June 2 to July 2.
Rates: rare if ever.

Phi Sagittariids (also Scorpiids–Sagittariids)
Maximum: June 18/19.
Lasts from: June 1 to July 15.
Rates: 5 per hour at best; weak maximum.

Ophiuchids
Maximum: June 20/21.
Lasts from: May 19 to July 2.
Rates: 6 per hour.
Comments: average magnitude 3 or fainter, weakly active over many years. NB: this stream can produce fireballs.

Corvids
Maximum: June 27/28 (June 26).*
Lasts from: June 25 to July 3 (June 25 to June 30, 1937).*
Rates: None reported since 1937.
Velocity: very slow.
Comments: could be related to asteroid 1979 VA, less likely Comet Tempel–Swift.

June Scutids
Maximum: June 27/28.
Lasts from: June 2 to July 29.
Rates: 2–4 per hour.

June Bootids
Maximum: June 28/29.
Lasts from: June 27 to July 5 (just June 28).*
Rates: 1–2 per hour.
Velocity: very slow.
Comments: strong in 1916, 1921, and 1927. Average magnitude is 5, though bright meteors do appear. Probable association with C/Pons–Winnecke. In 1969, Zdenek Sekanina detected related streams that he called Alpha Draconids and Bootids–Draconids.

July Pegasids

Maximum: early July.

Lasts from: July 1 to July 15.

Rates: 4 per hour.

Comments: I have seen meteors from a radiant apparently in Pegasus in the late 1980s. It is listed as my 99th meteor shower observation in my log. In July 1998, a group from Japan including Mitsue Sakaguchi recorded meteors from a Pegasus radiant.[4] On July 1/2, 2006, from New Plymouth, New Zealand, Rodney Austin and I observed several meteors emanating from a radiant in Pegasus. Possible photographic meteor recorded July 8, 2006.

Theta Aquilids[+]

Maximum: July 3.

Lasts from: June 8 to July 20.

Velocity: 40 km/s.[+]

Sigma Cassiopeids[+]

Maximum: July 7.

Lasts from: July 5 to July 13.

Rates: very low.

Velocity: 40 km/s.[+]

Vulpeculids[+]

Maximum: July 8.

Lasts from: July 7 to July 9.

Rates: very low.

Velocity: 32 km/s.[+]

Beta Equulids

Maximum: July 9.

Lasts from: July 7 to July 15.

Velocity: 34 km/s.[+]

Sigma Capricornids

Maximum: July 10–20.

Lasts from: June 18 to July 30.

Tau Capricornids

Maximum: July 12/13.

---

[4] Mitsue Sakaguchi *et al.* www.din.or.jp/~thashi/Inf1998_07_JP_E.htm (Minor Meteor Shower Circular, T. Hashimoto, webmaster).

Lasts from: June 2 to July 29.
Comments: orbit related to Northern Iota Aquariids.

Alpha Ursa Minorids
Maximum: July 13/14.
Lasts from: August 9 to August 30.
Rates: 1–4 per hour.
Comments: may be stronger as a telescopic shower. Discovered by W.A. Feibelman in 1961. Visual meteors are very rare; mostly radar detections.

July Phoenicids
Maximum: July 14.*
Lasts from: July 3 to July 18,* July 9 to July 17.
Rates: 1 per hour visually.
Velocity: 47 ± 3 km/s.
Comments: discovered by A.A. Weiss, 1957. Since radiant is at −47 degrees, this shower is not visible from most of the northern hemisphere.

Alpha Lyrids
Maximum: July 14/15.
Lasts from: July 9 to July 20.
Velocity: fast.
Comments: radiant motion to the southeast. See also Chapter 10.

Omicron Draconids
Maximum: July 16,* July 17/18.
Lasts from: July 7 to July 24,* July 6 to July 28.
Rates: low.
Velocity: 23.6 km/s.*
Comments: relation to Comet Metcalf C/1919 Q2. See Chapter 10.

Kappa Cassiopeids[+]
Maximum: July 20.
Lasts from: July 13 to July 28.
Velocity: 45 km/s.[+]

Southern Delta Aquarids
Maximum: July 29.
Lasts from: July 21 to August 29,* July 14 to August 18.
Rates: 15–20 per hour.
Velocity: 41.4 km/s,* 43 km/s.[+]
Comments: stronger of the two Delta Aquarid showers. See Chapter 11.

Alpha Capricornids
Maximum: July 30,* August 1/2.
Lasts from: July 15 to August 10,* July 15 to September 11.
Rates: 6–14 per hour.
Velocity: 22.8 km/s,* 25 km/s.[+]

Alpha Piscids Australids
Maximum: July 30/31.
Lasts from: July 16 to August 13.
Rates: 3–5 per hour.
Velocity: 46 km/s.[+]

Southern Iota Aquarids
Maximum: August 5,* August 6/7.
Lasts from: July 15 to August 25,* July 1 to September 18.
Rates: part of Aquarid complex.
Velocity: 33.8 km/s,* 32 km/s.[+]
Comments: brighter meteors than northern stream.

Upsilon Pegasids
Maximum: August 8/9.
Lasts from: July 25 to August 19.
Rates: 0–2 per hour on average.
Comments: with the sighting of three meteors on August 8, 1975, and more on subsequent nights, Harold Povenmire discovered evidence for this shower. Its maximum has been observed to occur a few days before the Perseid maximum, and weakens rapidly after that. Although the meteors are generally faint, there was one major fireball, on August 19, 1982.

Perseids
Maximum: August 12/13.
Lasts from: July 23 to August 22 (23).*
Rates: 50 or more per hour.
Velocity: 59.4 km/s.*
Comments: see Chapter 12.

Northern Delta Aquarids
Maximum: August 12,* August 13/14.
Lasts from: July 14 to August 25,* July 16 to September 10.
Rates: 10 per hour.
Velocity: 42.3 km/s,* 39 km/s.[+]
Comments: weaker of the two Delta Aquarid showers. See Chapter 11.

August Eridanids

Maximum: August 11/12.

Lasts from: August 2 to August 27.

Rates: radio detections.

Comments: discovered by Gary Kronk while studying radar meteor orbits. Possible connection to Periodic Comet Pons–Gambart (D/1827 M1).

Kappa Cygnids

Maximum: August 18.

Lasts from: July 26 to September 1 (August 9 to October 6).*

Rates: 5–6 per hour.

Velocity: 24.8 km/s.

Comments: this shower has been known for more than 150 years. Recently a fascinating paper by Daniel Jones, Iwan Williams, and Vladimir Porubcan has suggested that the Kappa Cygnid shower is a complex of showers with similar orbits. Their study noted similarity of orbits of meteors from five substreams, including the Alpha Lyrids, and has found a relation. Also, asteroids 2001 MG1 and 2004 LA12 may be fragments of the parent of the complex.[5]

Northern Iota Aquarids

Maximum: August 20,* August 25/26.

Lasts from: July 15 to September 20,* August 11 to September 10.

Rates: part of Aquarid complex.

Velocity: 31.2 km/s,* 31 km/s.[+]

August Pavonids

Maximum: August 31.

Velocity: 18.7 km/s.

Comments: in 1991 Peter Brown reported that Comet Levy (P/1991 L3), a Halley-type comet, is a potential source of meteoroids. The radiant is at 21 27.6 (321.9 degrees) and −62.1 degrees. Meteors were reported from that radiant in late August 2006. See Chapter 13.

Alpha Aurigids

Maximum: September 1.*

Lasts from: August 25 to September 6.

Rates: varies from 9 to 30 per hour.

Velocity: 66.3 km/s.

Comments: probable association with Comet Kiess (C/1911N1). Usually this

---

[5] D. C. Jones, I. Williams, and V. Porubcan, The Kappa Cygnid meteoroid complex, *Monthly Notices of the Royal Astronomical Society*, June 2006, **371** (2), 687.

shower is very weak, but in 1935 and 1986 hourly rates exceeding 20 were observed. Meteor astronomer Peter Jenniskens predicted storm-strength levels for the shower in 2007.

Gamma Aquarids
Maximum: September 7/8.
Lasts from: September 1 to September 14.
Rates: 1–4 per hour.

Alpha Triangulids
Maximum: September 11/12.
Lasts from: September 5 to 15, maybe longer.
Velocity: slow to medium.
Comments: discovered by Gary Kronk and Kurt Sleeter, and by George Gliba, in September 1993.

Eta Draconids
Maximum: September 12/13.
Lasts from: August 28 to September 23.
Rates: low.
Comments: best observed in early evening hours.

Southern Piscids
Maximum: September 11–20.
Lasts from: August 12 to October 7.
Rates: 5 per hour.
Comments: note the long duration of this weak shower.

Kappa Aquarids
Maximum: September 20/21.
Lasts from: September 11 to September 28.
Rates: low.
Velocity: 18.7 km/s.

Gamma Piscids
Maximum: September 23/24.
Lasts from: August 26 to October 22.
Rates: virtually none visually.
Comments: discovered by Zdenek Sekanina of the Jet Propulsion Laboratory from data from the Radio Meteor Project during the 1960s.

Annual Andromedids*
Maximum: October 3.*

Lasts from: September 25 to November 2.

Velocity: 23.2 km/s.

Eta Cetids

Maximum: October 1–5.

Lasts from: September 20 to October 2.

Rates: virtually none visually.

Comments: very rare, but a −20 fireball was recorded on October 9, 1969.

October Cygnids

Maximum: October 4–9.

Lasts from: September 22 to October 11.

Rates: very low.

Comments: recorded variously as Delta Cygnids and Alpha Cygnids.

October Cetids

Maximum: October 5/6.

Lasts from: September 8 to October 30.

Rates: low.

Velocity: fast.

Delta Aurigids

Maximum: October 6–15.

Lasts from: September 22 to October 23.

Rates: low.

Comments: discovered by Jack Drummond, Herbert Beebe, and Robert Hill at New Mexico State University. Visual surveys the following year yielded a small number of meteors.

Autumn Arietids

Maximum: October 8/9.

Lasts from: September 7 to October 27.

Rates: low.

Comments: part of a group of small radiants in the region of Aries and Taurus.

October Draconids

Maximum: October 9/10 (October 9).*

Lasts from: October 6 to October 10 (October 9).*

Rates: variable.

Velocity: 23 km/s.[+]

Comments: first detection of a meteor by radar. This shower is related to Comet Giacobini–Zinner. On October 9, 1946, a storm resulted when Earth crossed the path of debris from Comet Giacobini–Zinner which had crossed the Earth's orbit less than three weeks earlier. The annual shower is called the Draconids, but is nicknamed the "Giacobinids" occasionally after its parent comet. For five hours on that memorable night, and in spite of a bright Moon, some Canadian observers under Isabel K. Williamson counted over 2000 meteors. Since they observed only through specially made rings that allowed viewing of only selected areas, the count was far below what the actual total was.

Northern Piscids
Maximum: October 12/13.
Lasts from: October 5 to October 16 (September 25 to October 10).*
Rates: 5 per hour at maximum.
Velocity: 29 km/s.

Orionids
Maximum: October 20/21.
Lasts from: October 15 to October 29.
Rates: variable.
Velocity: 67 km/s.[+]
Comments: See Chapter 14.

Epsilon Geminids
Maximum: October 18/19 (October 19).*
Lasts from: October 10 to October 27 (October 14 to October 27).*
Rates: 1–2 per hour.
Velocity: 69.4 km/s.

Leo Minorids*
Maximum: October 24.*
Lasts from: October 22 to October 24.
Rates: 1–2 per hour.
Velocity: 61.8 km/s.

Alpha Pegasids
Maximum: November 1–12 (November 12).*
Lasts from: October 9 to November 17 (October 29 to November 2).*
Rates: very low, if any.
Velocity: 11.2 km/s; very slow.
Comments: possible association the lost Periodic Comet Blanpain (1819 IV).

Northern Taurids
Maximum: November 4–7.
Lasts from: October 12 to December 2.

Velocity: 30 km/s.[+]
Comments: See Chapter 15.

Southern Taurids
Maximum: October 30–November 7.
Lasts from: September 17 to November 27.
Velocity: 30 km/s.[+]
Comments: See Chapter 15.

Andromedids, or Bielids
Maximum: November 14/15.
Lasts from: September 26 to December 6.
Rates: from storm strength to nonexistent.
Comments: this shower comes from the lost Periodic Comet Biela. In 1846 this comet split. In 1852, observations of a brighter and a fainter component were made, and the comet has not been seen since. The associated meteor shower has occurred on different dates over the years. December 6, 1798, was the first heavy shower. Because the comet orbit's ascending node was decreasing, the shower's maximum was getting earlier and earlier. The story of the famous Stonyhurst observations of that shower on November 27, 1872, is told in Chapter 3.

Leonids
Maximum: November 17–19.
Lasts from: November 13 to November 20 (November 14 to November 20).
Rates: several per hour; thousands during storm years.
Velocity: 70.7 km/s,[*] 70 km/s.[+]
Comments: see Chapter 16.

Alpha Monocerotids
Maximum: November 21.
Lasts from: November 13 to December 2.
Rates: a few per hour but brief outbursts of up to 100 per hour every 10 years.
Comments: possible association with Comet van Gent–Peltier–Daimaca (1944 I).

Nu Orionids[+]
Maximum: November 28.
Lasts from: November 13 to December 5.
Velocity: 45 km/s.[+]

Alpha Puppids
Maximum: December 2–5.

Lasts from: November 17 to December 9.

Rates: up to 7 per hour.

Phoenecids

Maximum: December 5/6.

Lasts from: November 29 to December 9.

Rates: 1–5 per hour.

Comments: in 1956, the year of this shower's discovery, the rate was about 100 per hour. Probable connection to Comet Blanpain (1819 IV). The comet was rediscovered recently and is now known as P/Blanpain–Catalina (2003 WY25).

Delta Arietids

Maximum: December 8/9.

Lasts from: December 8 to January 2 (December 8 to December 14).*

Rates: very low.

Velocity: 13.2 km/s (very slow).

Comments: bright fireballs.

11 Canis Minorids

Maximum: December 10/11.

Lasts from: December 4 to December 15.

Comments: discovered by Keith Hindley as he was watching telescopic Geminids. He found five meteors whose paths intersected near the star 11 Canis Minoris. There is a possible connection between Comet Mellish (D/1917 F1), this shower, and the December Monocerotids.

Chi Orionids

Maximum: December 10/11 (northern: December 10; southern: December 11).*

Lasts from: N branch: November 16 to December 16 (December 4 to December 15).*

S branch: December 2 to December 18 (December 7 to December 14).*

Rates: 2–3 per hour.

Velocity: slow.

Comments: meteors can be bright, but note possible confusion with Geminids, which are also active at this time.

Sigma Hydrids

Maximum: December 11/12.

Lasts from: December 4 to December 15 (December 3 to December 15).*

Rates: 3–5 per hour.

Velocity: 58.4 km/s.

Comments: one of many showers discovered during the Harvard Meteor Project during the 1950s and 1960s.

December Monocerotids
Maximum: December 11/12 (December 10).*
Lasts from: November 12 to December 18 (November 27 to December 17).*
Velocity: 42.4 km/s,* 43 km/s.[+]
Comments: meteors mostly faint; however, this shower might be responsible for a series of bright fireballs recorded during the eleventh century according to Jones *et al.*[6]

Geminids
Maximum: December 13/14.
Lasts from: December 6 to December 19 (December 4 to December 16).*
Rates: up to 100 per hour.
Velocity: 34.4 km/s,* 37 km/s.[+]
Comments: see Chapter 17.

December Phoenicids
Maximum: December 5, 1956.
Lasts from: just recorded that one night.
Velocity: 21.7 km/s.

Coma Berenicids
Maximum: December 18–January 6.
Lasts from: December 8 to January 23 (December 12 to January 23).*
Rates: 1 or 2 per hour.
Velocity: fast, 65 km/s.
Comments: possible association with a never-confirmed comet discovery in 1912.

Ursids
Maximum: December 21/22.
Lasts from: December 17 to December 25.
Velocity: 33.4 km/s.*
Comments: see Chapter 18.

Sigma Serpentids[+]
Maximum: December 27.
Lasts from: December 13 to December 31.
Velocity: 44 km/s.[+]

---

[6] D.C. Jones *et al.*, The Kappa Cygnid meteoroid complex.

December Cepheids

Maximum: December 31/January 1.

Lasts from: December 31 to January 2.

Velocity: very slow, 13.9 km/s.

Comments: unconfirmed. Meteors from Comet Levy P/2006 T1. I observed on the night before maximum and saw possibly one Cepheid meteor, about magnitude 5, slow moving. The morning of maximum, the sky was covered by a layer of cirrostratus clouds.

Appendix

Meteor societies and books

The International Meteor Organization. Founded in 1988, the IMO seeks to bring together meteor enthusiasts from all over the world. www.imo.net.

The American Meteor Society. Founded in 1911, this organization brings together serious meteor observers. www.amsmeteors.org.

Royal Astronomical Society of New Zealand. For southern hemi-sphere observers, the RASNZ has an active comet and meteor section headed by John Drummond. www.rasnz.org.nz/Sections.htm#comet.

The Dutch Meteor Society is one of the most active of the European meteor organizations. home.wxs.nl/~terkuile/.

Another excellent website is Gary Kronk's Meteor Calendar. It offers detailed histories and observing information for major and minor showers each month. comets.amsmeteors.org/meteors/calendar.html.

In addition, almost every national astronomy society has a section, or a program, involving meteor observing. These sections can usually be found in the web pages of the various national societies.

There are several good books on meteors and meteorites. Mike Reynolds' *Falling Stars: A Guide to Meteors and Meteorites* (Mechanicsburg, PA: Stackpole Books, 2001) offers advice from an experienced meteorite collector and meteor observer.

Finally, more advanced readers will enjoy and benefit from Peter Jenniskens' *Meteor Showers and their Parent Comets* (Cambridge: Cambridge University Press, 2006).

Index